MW00700031

ESSENTIALS OF ANATOMY AND PHYSIOLOGY LABORATORY MANUAL

ESSENTIALS OF ANATOMY AND PHYSIOLOGY LABORATORY MANUAL

CONNIE ALLEN
VALERIE HARPER

Edison Community College

John Wiley & Sons, Inc

Executive Editor	Bonnie Roesch
Project Editor	Mary O'Sullivan
Production Manager	Pamela Kennedy
Production Editor	Sarah Wolfman-Robichaud
Senior Marketing Manager	Clay Stone
Senior Designer	Kevin Murphy
Photo Manager	Hilary Newman
Illustration Editor	Anna Melhorn
Production Management Services	Carole Kuhn/GTS Companies
Cover Photos	Skeletal art: Leonard Dank
	Cardiac muscle tissue: © Ed Reschke

This book was typeset in 10/12 Times Roman by GTS Companies and printed and bound by Courier Westford Press. The cover was printed by Lehigh Press.

The paper in this book was manufactured by a mill whose forest management programs include sustained yield harvesting of its timberlands. Sustained yield harvesting principles ensure that the number of trees cut each year does not exceed the amount of new growth.

The procedures in this text are intended for use only by students with appropriate faculty supervision. In preparing the text, care has been taken to identify potentially hazardous steps and to insert safety precautions where appropriate. The authors and publisher believe the procedures to be useful tools if performed with the materials and equipment specified, in careful accordance with the instructions and methods in the text. However, these procedures must be conducted at one's own risk. The author and publisher do not warrant or guarantee the safety of individuals using these procedures and specifically disclaim any and all liability resulting directly or indirectly from the use or application of any information contained in this book.

This book is printed on acid-free paper. ∞

Copyright © 2004 by John Wiley & Sons, Inc. All rights reserved.

No part of this publication may be reproduced, stored in a retrieval system, or transmitted in any form or by any means, electronic, mechanical, photocopying, recording, scanning, or otherwise, except as permitted under Section 107 or 108 of the 1976 United States Copyright Act, without either the prior written permission of the publisher or authorization through payment of the appropriate per-copy fee to the Copyright Clearance Center, 222 Rosewood Drive, Danvers, MA 01923, (978) 750-8400, fax (978) 750-4470. Requests to the Publisher for permission should be addressed to the Permissions Department, John Wiley & Sons, Inc., 111 River Street, Hoboken, NJ 07030 (201) 748-6008, E-mail: PERMREQ@WILEY.COM. To order books or for customer service, call 1-800-CALL-WILEY (225-5945).

Printed in the United States of America.

I dedicate this book to my husband Jim, and my children,
Burke and Staci, Brittany and Michael, and Michaela,
for their unwavering love, patience, and encouragement.
—CONNIE ALLEN

I dedicate this book to my husband Chuck, and our children,
Scott and Kate, for their love, support, and patience.
—VALERIE HARPER

We also want to dedicate this book to a special colleague and friend,
Sandra Grabowski, who was always eager to encourage and support us.
We will miss her gentle spirit and loving heart.

PREFACE

Anatomy and physiology is a challenging course, and this laboratory manual is written to help students meet that challenge. It is written for students interested in allied health fields who are enrolled in a one-semester course. Although this manual incorporates many illustrations from *Introduction to the Human Body, 6th edition*, by Gerard J. Tortora and Sandra R. Grabowski, it may be used with any one-semester anatomy and physiology textbook.

The design of this laboratory manual is based on the authors' experience as anatomy and physiology instructors and uses three learning styles: visual, auditory, and kinesthetic. When students label diagrams, they focus on the structure rather than just the dot at the end of a line. Writing out the structure's name and pronouncing it reinforces learning. Also, having students become subjects of laboratory exercises makes learning personal. Animal dissections give students an opportunity to physically manipulate structures, comparing location and texture, and to observe how structures are supported, protected, and attached by connective tissue.

Special features incorporated in this laboratory manual include:

- Just enough text is provided to introduce concepts in each section and to set up and support the laboratory section. The exercises are written so that students do not need their textbooks to complete the laboratory activities.
- Unlabeled drawings, photographs, and photomicrographs are included for students to label either at home or in the laboratory. Students first write out the name of the structure to help memorize it. Then the completed diagrams will be used to identify structures on models.
- An answer key is provided at the back of the laboratory manual for the students to self-correct their diagrams.
- Word derivatives for bolded terms are given in the text.
- Phonetic pronunciation is included as new words are introduced. Students are directed to pronounce

the name of a structure as they point to it on a model. Most students have difficulty pronouncing the anatomical terms and need practice saying and hearing the words.

- Physiology experiments use students as subjects and can be completed with either simple, inexpensive equipment or more complex setups.
- Discussion Questions are within the activities to make the students think about the material presented.
- An Answer Key is provided at the end of the laboratory manual for the activities in each exercise. Students receive immediate feedback, and they are not dependent on the instructor for the correct answers.
- A "Reviewing Your Knowledge" section follows the activities at the end of each exercise. This section provides a thorough review of the material in the exercise. This section may be handed in to the instructor for a grade because the answers are not in the back of the laboratory manual. Answers to this section is in the online Instructor's Manual.
- Each exercise contains a list of objectives, materials needed for the exercise, and easily identifiable laboratory activity sections.

Supplemental Materials

- Cat Dissection Manual. Contains dissection activities for use in the lab accompanied by full color photos and figures.
- Fetal Pig Dissection Manual. Contains dissection activities for use in the lab accompanied by full color photos and figures.
- Instructor's Resources. Available on the book companion web site at www.wiley.com/college/allen. Contains an Answer Key for "Reviewing Your Knowledge."
- Student Resources. Available on the book companion web site at www.wiley.com/college/allen. Contains Anatomy Drill and Practice and Cadaver Practical Labeling exercises.

ACKNOWLEDGEMENTS

We deeply appreciate the support, instruction, and encouragement from the members of our editorial, production, and marketing team at Wiley: Bonnie Roesch, Executive Editor, Mary O'Sullivan, Project Editor, Sarah Wolfman-Robichaud, Production Editor, Hilary Newman, Photo Manager, Anna Melhorn, Illustration Editor, and Clay Stone, Marketing Manager. We also wish to thank Gerard Tortora and Sandra Grabowski for producing a wonderful textbook that provided many illustrations and ideas for our laboratory manual.

Reviewers

A special thank you to all the reviewers. Their suggestions were extremely helpful in developing this laboratory manual.

Tom Allen, Lake Superior State University, Mark Bloom, Tyler Junior College, LuAnne Clark, Lansing Community College, James Crowder, Brookdale Community College, Connie Dempsey, Stark State College, Philip Eichman, University of Rio Grande, Julia Huggins, Arkansas State University, Roy Hyle, Thomas Nelson Community College, John Harley, Eastern Kentucky University, Murray Jensen, University of Minnesota, Donald Linzey, Wytheville Community College, Dan McCann, Gonzaga University, Steven Nunez, Saulk Valley Community College, Amy Ouchley, University of Louisiana Monroe, Barbara Rendleman, Ivy Tech State College SW, Marilyn Shopper, Johnson County Community College, Rick Simonson, University of Nebraska at Kearney, Caryl Tickner, Stark State College, Betty Weiss, Emmanuel College, Claudia Williams, Campbell University.

CONTENTS

ANATOMICAL LANGUAGE

After completing this exercise, you should be able to:

- Describe the anatomical position
- Identify major body regions
- Use anatomical terms correctly
- Use directional terms correctly
- Identify body planes and sections

- human models or anatomical charts
- articulated skeleton
- 5 sheep brains (for class demonstration)
- apples (1 per group) and plastic knives
- plastic tubing (1 foot piece per group)

Students often complain that they are not just learning anatomy and physiology, but are learning a new language. This is quite true. In this exercise, you will learn anatomical language that will be used throughout the course and also to communicate effectively with other health care professionals.

A. ANATOMICAL POSITION

The anatomical position is the reference position anatomists and people in the medical field use to describe the location of body parts or regions. In the anatomical position, the body is erect and facing forward; the arms are straight and at the sides of the body with the palms facing forward; the legs are straight with the feet facing forward and flat (Figure 1.1).

ACTIVITY 1 ANATOMICAL POSITION

1 Assume the anatomical position.

2 Have your laboratory partner determine if you are in the correct position.

B. BODY REGIONS

The main body regions are described in Table 1.1. It is important that you learn the correct boundaries for each region. Two common misconceptions are that the arm is the area between the shoulder and wrist and that the leg includes the thigh. The arm is located between the shoulder and elbow, and the forearm is located from the elbow to wrist. The thigh is located between the groin and knee, and the leg is located between the knee and ankle.

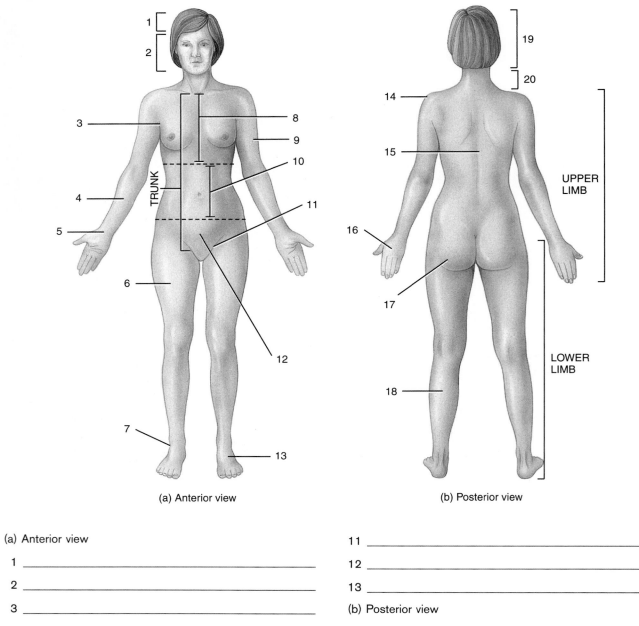

(a) Anterior view

(b) Posterior view

(a) Anterior view

1 _____

2 _____

3 _____

4 _____

5 _____

6 _____

7 _____

8 _____

9 _____

10 _____

11 _____

12 _____

13 _____

(b) Posterior view

14 _____

15 _____

16 _____

17 _____

18 _____

19 _____

20 _____

FIGURE 1.1 **Body regions.**

TABLE 1.1 Body Regions

BODY REGION	DESCRIPTION
HEAD	
Face	Portion of head not normally covered by scalp
Skull	Bony portion of head
NECK	Body area between head and trunk
TRUNK	Central body area to which head and limbs are attached
Chest	Area of trunk between neck and abdomen; contains heart and lungs; diaphragm forms boundary between chest and abdomen
Abdomen	Area of trunk between chest and pelvis; contains digestive organs; hip bones form lower boundary of abdomen
Pelvis	Area of trunk below abdomen; contains internal reproductive organs and urinary bladder
Back	Posterior portion of trunk between neck and buttocks
UPPER LIMB	
Shoulder	Curved area where upper limb attaches to upper border of trunk
Armpit	Underarm area where upper limb attaches to trunk
Arm	Area of upper limb between shoulder and elbow
Forearm	Area of upper limb between elbow and wrist
Wrist	Portion of hand that connects hand to forearm
Hand	Includes wrist and fingers
LOWER LIMB	
Buttocks	Rounded area on posterior (back) surface where lower limb attaches to trunk
Groin	Area on the anterior (front) surface where lower limb attaches to the pelvis
Thigh	Area of lower limb above the knee
Leg	Area of lower limb between knee and ankle
Ankle	Portion of foot that connects foot to leg
Foot	Includes ankle and toes

ACTIVITY 2 BODY REGIONS

1 Label Figure 1.1 with the appropriate body regions. Use terms from Table 1.1.

2 Identify body regions on models or anatomical charts. Use terms from Table 1.1.

3 Pronounce the terms as you point to them.

C. ANATOMICAL TERMS

Anatomical terms describe body regions, specific body areas, and landmarks. These terms are often part of the names of muscles, bones, nerves, and blood vessels. Most of these words are derived from Latin or Greek. Many words have one or more word roots with a prefix and/or a suffix added. For example, in the word *antecubital, ante-* is a prefix meaning before or in front of; the word root *cubit-* means elbow; *-al* is a suffix meaning pertaining to. Table 1.2 contains anatomical terms with four different suffixes that all mean "pertaining to." These suffixes are: *-al, -ic, -ar,* and *-ary*. When suffixes like these are added to word roots, they form an adjective. Nouns have different endings, such as *-um, -us, -is,* or *-a*. For example, sternum is a bone (noun) and sternal is an adjective. Learning these terms at this time will help you throughout the course. Anatomical terms and their definitions are found in Table 1.2.

ACTIVITY 3 ANATOMICAL TERMS

1 Label Figure 1.2 with the appropriate anatomical terms. Use terms from Table 1.2.

2 Identify the specific body regions, areas, and landmarks on models or anatomical charts.

3 Pronounce the anatomical terms as you point to them on models or anatomical charts.

TABLE 1.2 Anatomical Terms

TERM	DEFINITION	TERM	DEFINITION
Axial	Pertaining to the central part of the body, the head and trunk	**APPENDICULAR**	Pertaining to the extremities or limbs
Cephalic (se-FAL-ik)	Pertaining to the head	***Upper Appendage***	Upper limb
• Facial	Pertaining to the face	• Acromial (a-KROM-ee-al)	Pertaining to the highest point of the shoulder
• Frontal	Pertaining to the forehead		
• Orbital	Pertaining to the eye	• Axillary	Pertaining to the armpit
• Otic (OH-tik)	Pertaining to the ear	• Brachial (BRAY-key-ul)	Pertaining to the arm
• Nasal	Pertaining to the nose	• Antecubital	Pertaining to the anterior (front) surface of the elbow
• Buccal (BUCK-al)	Pertaining to the cheek		
• Oral	Pertaining to the mouth	• Olecranal (oh-LEK-ra-nul)	Pertaining to the posterior (back) surface of the elbow
• Cranial	Pertaining to the skull		
• Occipital (ox-SIP-i-tal)	Pertaining to the back of head	• Antebrachial	Pertaining to the forearm
Cervical	Pertaining to the neck	• Carpal	Pertaining to the wrist
		• Digital	Pertaining to the digits (fingers)
Thoracic	Pertaining to the chest		
• Pectoral (PEK-tore-al)	Pertaining to the breast area	***Lower Appendage***	Lower limb
• Sternal	Pertaining to the breast bone	• Inguinal (ING-won-ul)	Pertaining to the groin where the thigh attaches to the pelvis
Abdominal	Pertaining to the abdomen		
• Ventral	Pertaining to the belly		
• Umbilical	Pertaining to the navel	• Gluteal	Pertaining to the buttocks
• Coxal (COX-al)	Pertaining to the hip	• Femoral	Pertaining to the thigh
Pelvic	Pertaining to the pelvis	• Patellar	Pertaining to the anterior (front) surface of the knee
• Pubic	Pertaining to the genital area		
		• Popliteal (pop-lih-TEE-ul)	Pertaining to the posterior (back) surface of the knee
Dorsal	Pertaining to the back		
• Scapular	Pertaining to the shoulder blade region	• Fibular or peroneal	Pertaining to the lateral side of the leg
		• Tarsal	Pertaining to the ankle
• Vertebral	Pertaining to the spine	• Pedal	Pertaining to the foot
• Lumbar	Pertaining to the lower back or loin region	• Calcaneal (kal-KANE-ee-ul)	Pertaining to the heel
		• Plantar	Pertaining to the sole of foot
		• Digital	Pertaining to the digits (toes)

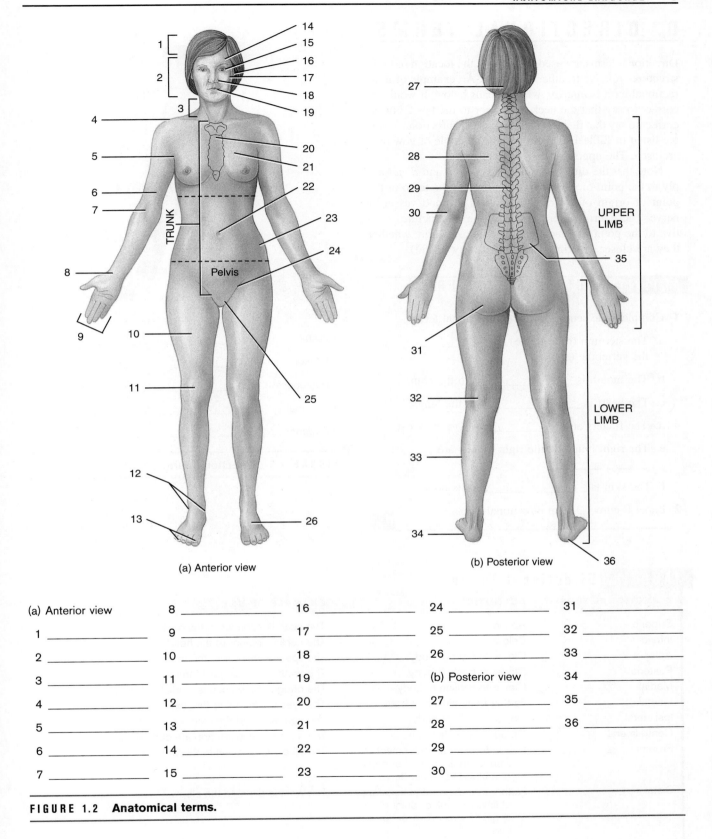

(a) Anterior view

(b) Posterior view

(a) Anterior view	8 _____	16 _____	24 _____	31 _____
1 _____	9 _____	17 _____	25 _____	32 _____
2 _____	10 _____	18 _____	26 _____	33 _____
3 _____	11 _____	19 _____	(b) Posterior view	34 _____
4 _____	12 _____	20 _____	27 _____	35 _____
5 _____	13 _____	21 _____	28 _____	36 _____
6 _____	14 _____	22 _____	29 _____	
7 _____	15 _____	23 _____	30 _____	

FIGURE 1.2 **Anatomical terms.**

D. DIRECTIONAL TERMS

Directional terms are used to describe the location of body structures relative to other structures. An example of a directional term is *inferior,* which means below. It would be correct to say that the neck is inferior to the head, but incorrect to say that the neck is inferior. The directional terms are listed in Table 1.3, along with an example of how they are used. The opposite terms are paired.

Note that the directional terms *proximal* and *distal* apply to the point of attachment of a limb to the torso or the point of origin of a structure such as a blood vessel or nerve. These terms refer to the location of structures relative to the point of attachment or point of origin, whether they are closer (proximal) or farther away (distal).

ACTIVITY 4 DIRECTIONAL TERMS

1 Complete the sentences using directional terms.

   a. The sternum (breastplate) is _____ to the vertebrae (spine).

   b. The mouth is _____ to the chin.

   c. The heart is _____ to the lungs.

   d. The fingers are _____ to the wrist.

   e. The right lung and the right kidney are

   _____.

   f. The skin is _____ to the bones.

2 Label Figure 1.3 with directional terms.

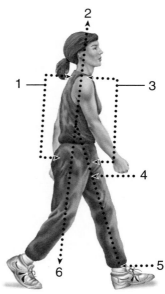

- Anterior   1 _____

- Distal   2 _____

- Inferior   3 _____

- Posterior   4 _____

- Proximal   5 _____

- Superior   6 _____

FIGURE 1.3   Directional terms.

TABLE 1.3 Directional Terms

DIRECTIONAL TERM	DEFINITION	EXAMPLE OF USE
Superior	Above	The head is superior to the neck.
Inferior	Below	The neck is inferior to the head.
Anterior	Closer to front of body	The lips are anterior to the teeth.
Posterior	Closer to back of body	The teeth are posterior to the lips.
Medial	Closer to midline of body	The bridge of the nose is medial to the eyes.
Lateral	Farther from midline of body	The eyes are lateral to the nose.
Ipsilateral	On same side of body	The right arm and right leg are ipsilateral.
Contralateral	On opposite sides of body	The right arm and left arm are contralateral.
Proximal	Nearer to point of attachment of limb to trunk or point of origin of a blood vessel or nerve	The elbow is proximal to the wrist.
Distal	Farther from point of attachment of limb to trunk or point of origin of a blood vessel or nerve	The wrist is distal to the elbow.
Superficial	Closer to surface of body	The skin is superficial to muscles.
Deep	Farther from surface of body	The muscles are deep to the skin.

E. BODY PLANES AND SECTIONS

Planes divide the body or organs into sections to observe internal structures (Figure 1.4). **Sagittal** (*sagitta* = arrow) **planes** pass vertically through the body or organs, and divide them into right and left sections (**sagittal sections**). A **midsagittal** plane passes vertically through the midline and divides the body into right and left halves. A **frontal** or **coronal plane** passes vertically through the body or organs and produces anterior and posterior sections (**frontal sections**). A **transverse plane** passes horizontally through the body and produces superior and inferior sections (**transverse sections**). Transverse sections are also called cross-sections. **Oblique planes** pass through the body at an angle. The type of plane will determine which type of sections are obtained and the structures that can be viewed in the sections.

Often, we look at sections of organs and tissues to observe structures. Sections that are produced by a plane running along the long axis of a long narrow structure, such as a blood vessel, are called **longitudinal sections**. Sections that are produced by a plane running perpendicular to the long axis are called **cross-sections**.

CLINICAL NOTE: *Transverse sections observed with computed tomography (CT) scans or magnetic resonance imaging (MRIs) are called axial sections.*

ACTIVITY 5 PLANES AND SECTIONS

1 Label the planes in Figure 1.4(a) and (b).

2 Cut sagittal, frontal, and transverse planes through an apple.

- Working in a group, draw a face on the apple.
- Cut sagittal, frontal, and transverse planes through the apple to make sagittal, frontal, and transverse sections.
- Observe the appearance of the apple core in each section.
- Keep sections together to form a whole apple to show to your instructor.

3 Observe how one plane can produce longitudinal and transverse sections through a tube.

- Tube-like structures are found throughout the body. Examples of these are blood vessels and kidney tubules that twist and bend as they travel through the body. A slice of tissue may contain a tube with

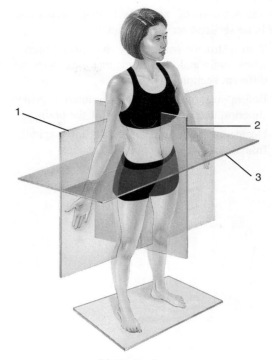

(a) Body planes

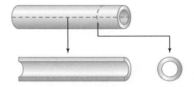

Longitudinal section   Cross-section

(b) Longitudinal and cross-sections

- Frontal   1 _____
- Sagittal   2 _____
- Transverse   3 _____

FIGURE 1.4 Body planes and sections.

longitudinal sections, cross-sections, and oblique sections.

- Observe a demonstration of a tube cut to display longitudinal sections and cross-sections.
- Twist a one-foot section of plastic tubing so that one plane would divide one area of the tube into a longitudinal section and another area into a cross-section.
- Do not cut the tube unless instructed to do so.
- Show your instructor where a cut would produce both a longitudinal section and a cross-section.

4 Demonstration of sagittal, frontal, transverse, and oblique sections on sheep brains.

- Your instructor will display five sheep brains: one whole brain and four brains that have been cut into different sections.

- Before you start, determine the anterior, posterior, superior, and inferior surfaces of the brain.

- Decide which brain has been cut into sagittal, frontal, transverse, or oblique sections.

Brain 1—Whole brain

Brain 2 _____ section

Brain 3 _____ section

Brain 4 _____ section

Brain 5 _____ section

- Compare the appearance of the different sections.

EXERCISE

1

REVIEWING YOUR KNOWLEDGE

A. Anatomical Position

The person in Figure 1.5 is not in anatomical position.
List four differences from anatomical position.

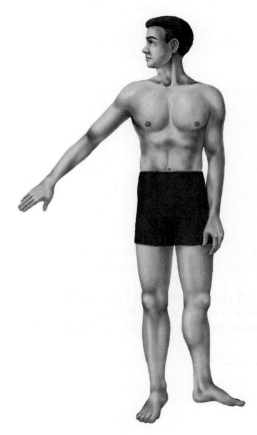

1. _____

2. _____

3. _____

4. _____

FIGURE 1.5   **Person not in anatomical position.**

B. Anatomical Terms

Write the anatomical terms that the phrase describes.

_____ **1.** Pertaining to the navel

_____ **2.** Pertaining to the area between neck and abdomen

_____ **3.** Pertaining to the ear

_____ **4.** Pertaining to the anterior surface of elbow region

_____ **5.** Pertaining to the face; anterior portion of head

_____ **6.** Pertaining to the nose

_____ **7.** Pertaining to the neck

_____ **8.** Pertaining to the posterior surface of knee

_____ **9.** Pertaining to the wrist

_____ **10.** Pertaining to the back

_____ **11.** Pertaining to the armpit area

_____ **12.** Pertaining to the mouth

_____ **13.** Pertaining to the anterior surface of knee

_____ **14.** Pertaining to the breastplate

_____ **15.** Pertaining to the hip

_____ **16.** Pertaining to the side of the leg

_____ **17.** Pertaining to the area between shoulder and elbow

_____ **18.** Pertaining to the fingers or toes

_____ **19.** Pertaining to the cheek

_____ **20.** Pertaining to the heel

_____ **21.** Pertaining to the sole of foot

_____ **22.** Pertaining to the groin where thigh attaches to pelvic region

_____ **23.** Pertaining to the head

_____ **24.** Pertaining to the foot

_____ **25.** Pertaining to the eye

_____ **26.** Pertaining to the genital area

_____ **27.** Pertaining to the area between hip and knee

_____ **28.** Pertaining to the area that includes bones enclosing the brain but excluding facial bones

_____ **29.** Pertaining to the forehead

_____ **30.** Pertaining to the spine

_____ **31.** Pertaining to the inferior back of the head

_____ **32.** Pertaining to the area of back between ribs and buttocks

_____ **33.** Pertaining to the belly; anterior area between diaphragm and pelvis

_____ **34.** Pertaining to the anterior trunk below the abdomen

_____ **35.** Pertaining to the area of back that contains shoulder blades

_____ **36.** Pertaining to the posterior surface of elbow region

C. Planes

Complete the sentences using the planes listed.

- Frontal
- Sagittal
- Transverse

1. The _____ plane divides body or organ into right and left sections.

2. The _____ plane divides body or organ into anterior and posterior sections.

3. The _____ plane divides body or organ into superior or inferior sections.

4. Which planes when passed through the body would result in two sections, with each section containing a piece of the heart and a piece of each lung?

    _____ and _____

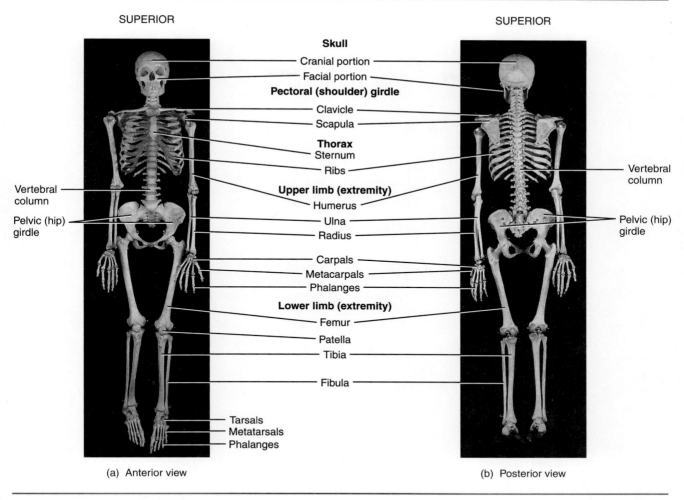

SUPERIOR

SUPERIOR

Skull
— Cranial portion —
— Facial portion —
Pectoral (shoulder) girdle
— Clavicle —
— Scapula —
Thorax
Sternum
— Ribs —
Upper limb (extremity)
— Humerus —
— Ulna —
— Radius —
— Carpals —
— Metacarpals —
— Phalanges —
Lower limb (extremity)
— Femur —
— Patella —
— Tibia —

— Fibula —

Tarsals
Metatarsals
Phalanges

Vertebral
column

Pelvic (hip)
girdle

Vertebral
column

Pelvic (hip)
girdle

(a) Anterior view

(b) Posterior view

FIGURE 1.6 **Human skeleton.**

D. Directional Terms

Complete the sentences using directional terms. Use Figure 1.6 for reference.

1. The clavicle is _____ to the ribs.

2. The ribs are _____ to the sternum.

3. The humerus is _____ to the radius.

4. The ulna is _____ to the radius.

5. The tibia is _____ to the femur.

6. The right humerus and the right radius are

_____.

7. The pelvic girdle is _____ to the ribs.

8. The sternum is _____ to the vertebral column.

9. The scapula is _____ to the clavicle.

10. The right fibula and the left fibula are

_____.

ORGAN SYSTEMS

OBJECTIVES

**After completing this exercise,
you should be able to:**

- Name the organ systems and describe their function
- Identify the major organs from each organ system
- Identify the body cavities and the organs they contain
- Identify the abdominopelvic quadrants and the organs found in each quadrant

MATERIALS

- torso models or charts
- male and female reproductive models or charts
- articulated skeleton

A. OVERVIEW OF ORGAN SYSTEMS

Organ systems work together to provide the cells of the body with an optimal environment for their needs. Maintenance of this optimal environment, within limits, is called **homeostasis** (*homeo-* = sameness; *-stasis* = standing). All organ systems cooperate to maintain homeostasis, and failure to maintain homeostasis results in disorders and disease. Organ system functions and major organs are summarized in Table 2.1.

ACTIVITY 1 IDENTIFYING ORGAN SYSTEMS

Identify the organ systems in Figure 2.1 using the organ system list provided.

TABLE 2.1 Functions and Major Organs of the Organ Systems

ORGAN SYSTEM	FUNCTION	MAJOR ORGANS
Cardiovascular	Transports nutrients, chemical messengers, gases, and wastes in blood	Heart, blood vessels
Respiratory	Adds oxygen to blood and removes carbon dioxide from blood; maintenance of carbon dioxide levels helps regulate pH	Nose, pharynx (throat), trachea, bronchi, lungs
Digestive	Breaks down food into units that can be absorbed into the body; eliminates solid wastes	Mouth, pharynx, esophagus, stomach, intestines, pancreas, liver, gall bladder
Urinary	Removes nitrogenous wastes; maintains body fluid volume, pH, and electrolyte levels through urine production	Kidneys, ureters, urinary bladder, urethra
Integumentary	Provides a protective barrier for the body, aids in production of vitamin D, and contains sensory receptors for pain, touch, and temperature	Skin, hair, nails, sweat glands, oil glands
Lymphatic and Immune	Returns fluid to cardiovascular system; detects, filters, and eliminates disease-causing organisms, including cancer cells	Lymph nodes, spleen, thymus
Skeletal	Protects major organs; provides levers and support for body movement	Bones
Muscular	Moves bones and maintains posture	Skeletal muscles
Nervous	Controls cell function with electrical signals called action potential	Brain, spinal cord, nerves
Endocrine	Controls cell function with chemical signals called hormones	Hypothalamus, pituitary, thyroid, thymus, adrenal glands, pancreas, ovaries, testes
Reproductive	Produces gametes: female uterus provides environment for development of fetus	Male: testes, penis, prostate gland Female: ovaries, uterus, vagina

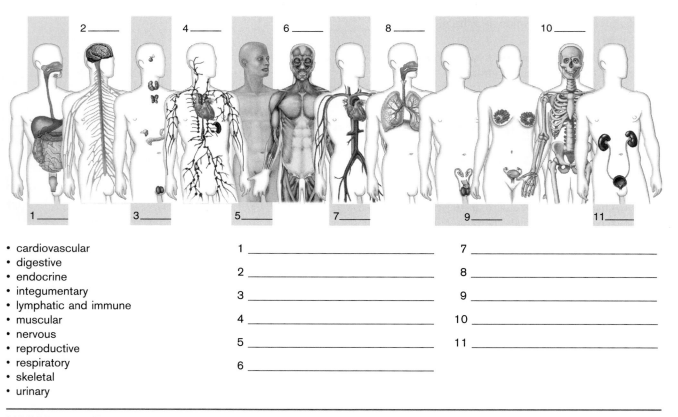

- cardiovascular
- digestive
- endocrine
- integumentary
- lymphatic and immune
- muscular
- nervous
- reproductive
- respiratory
- skeletal
- urinary

1 _____   7 _____

2 _____   8 _____

3 _____   9 _____

4 _____   10 _____

5 _____   11 _____

6 _____

FIGURE 2.1   Organ systems and selected organs.

B. IDENTIFICATION OF MAJOR ORGANS ON A TORSO MODEL

You will be identifying organs from anterior to posterior on a torso and answering questions concerning their position relative to the organs around them.

ACTIVITY 2 ORGAN SYSTEMS AND ORGANS

1 Identify the following organs on the anterior surface of a torso model. Identify all the organs without removing any organs from the model.
- brain
- heart
- large intestine
- liver (right side of torso)
- lungs
- small intestine
- stomach (left side of torso)
- trachea

2 Remove the lungs, heart, liver, and stomach. Identify the following organs.
- bronchi (right and left)
- esophagus
- gallbladder (inferior surface of liver)
- pancreas (posterior to stomach)
- spleen

3 Remove the large intestine and small intestine. Identify the following organs.
- abdominal aorta
- adrenal glands (superior surface of kidneys)
- appendix (lower right side of large intestine)
- inferior vena cava
- kidneys
- ureters
- urinary bladder

4 Identify the female reproductive organs on a female reproductive model or chart. Observe the position of the urinary bladder relative to the uterus.
- ovaries
- uterus

5 Identify the male reproductive organs on a male reproductive model or chart.
- penis
- scrotum (skin covering testes)
- testes

6 Answer the questions about the position of each organ.

1. The stomach is _____ to the small intestine.
   a. superior   b. inferior   c. medial   d. lateral

2. The liver is _____ to the lungs.
   a. superior   b. inferior   c. medial   d. lateral

3. The lungs are _____ to the heart.
   a. superior   b. inferior   c. medial   d. lateral

4. The trachea is _____ to the esophagus.
   a. superior   b. inferior   c. anterior   d. posterior

5. The pancreas is _____ to the stomach.
   a. medial   b. anterior   c. superior   d. posterior.

6. The gallbladder is on the _____ surface of the liver.
   a. superior   b. inferior   c. posterior   d. lateral

7. The stomach is _____ to the spleen.
   a. lateral   b. medial   c. superior   d. inferior

8. The abdominal aorta and inferior vena cava are _____ to the kidneys.
   a. medial   b. lateral   c. superior   d. inferior

9. The kidneys are _____ to the small intestine.
   a. anterior   b. posterior   c. superior   d. inferior

10. The urinary bladder is _____ to the uterus.
    a. posterior and superior
    b. anterior and inferior
    c. medial and superior
    d. lateral and posterior

C. BODY CAVITIES

The two main cavities of the body are the dorsal and ventral body cavities. The **dorsal body cavity** is located near the posterior surface of the body. It has two main subdivisions: the **cranial cavity** and the **vertebral** (*vertebra* = back) **canal** which are continuous. The **ventral body cavity** is located near the anterior surface of the body and has two subdivisions, the **thoracic cavity** and the **abdominopelvic cavity,** that are separated by the **diaphragm.** Many organs within the ventral body cavity are enclosed within serous membranes. Blood vessels, lymphatic vessels, and nerves are found in all body cavities.

1. Dorsal Body Cavity

Cranial cavity: Space within the skull that contains the brain.

Vertebral (spinal) canal: Space within vertebral column that contains the spinal cord.

2. Ventral Body Cavity

Thoracic cavity: Space enclosed by the ribs, sternum, and vertebral column. This cavity contains the lungs, bronchi, heart, trachea, esophagus, and thymus. The thoracic cavity contains three smaller cavities.

- One **pericardial cavity** (*peri-* = around; *-cardia* = heart): The heart is enclosed within a serous membrane called the **pericardium.** The space between the pericardium and the heart is the pericardial cavity.

- Two **pleural cavities** (*pleura* = rib): Each lung is enclosed within a serous membrane called the **pleura.** The space between each lung and the

pleural membrane that surrounds it is the pleural cavity.

Abdominopelvic cavity: Consists of two continuous cavities: the abdominal cavity and the pelvic cavity.

- **Abdominal cavity:** The superior portion of the abdominopelvic cavity that is located between the diaphragm superiorly and the brim of the pelvis inferiorly. This cavity contains the stomach, liver, gallbladder, pancreas, spleen, small intestine, adrenal glands, kidneys, part of the large intestine, and the appendix. Many organs within the abdominal cavity are enclosed within a serous membrane called the **peritoneum.**

- **Pelvic cavity:** The inferior portion of the abdominopelvic cavity located between the pelvic brim superiorly and the body wall inferiorly. This cavity contains the ovaries, uterus, vagina, prostate gland, urinary bladder, part of the large intestine, and the rectum. It is important to note that the testes are not located in the ventral body cavity but are outside the body wall.

ACTIVITY 3 BODY CAVITIES

1 Label the dorsal and ventral body cavities, their subdivisions, and the diaphragm on Figure 2.2(a) and Figure 2.2(b).

2 Locate the dorsal body cavity and its subdivisions on a skeleton.

3 Locate the ventral body cavity, its subdivisions, and the diaphragm on a torso model.

4 Complete Table 2.2 by listing the organ(s) found in each body cavity.

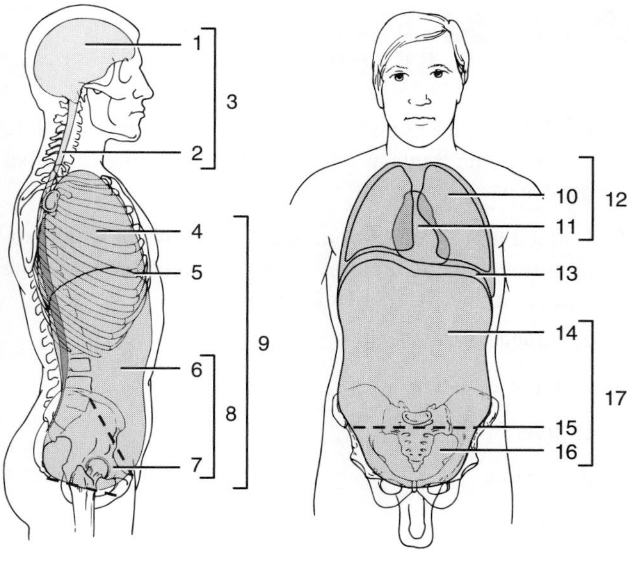

(a) Lateral view

(b) Anterior view

(a) Lateral view
- abdominal cavity
- abdominopelvic cavity
- cranial cavity
- diaphragm
- dorsal cavity
- pelvic cavity
- thoracic cavity
- ventral cavity
- vertebral canal

1 ____________
2 ____________
3 ____________
4 ____________
5 ____________
6 ____________
7 ____________
8 ____________
9 ____________

(b) Anterior view
- abdominal cavity
- abdominopelvic cavity
- brim of pelvis
- diaphragm
- pelvic cavity
- pericardial cavity
- pleural cavity
- thoracic cavity

10 ____________
11 ____________
12 ____________
13 ____________
14 ____________
15 ____________
16 ____________
17 ____________

FIGURE 2.2 Body cavities.

TABLE 2.2 Body Cavities and Organs

BODY CAVITY	ORGANS		
Dorsal body cavity			
Cranial cavity	1. ____________		
Vertebral canal	2. ____________		
Ventral body cavity			
Thoracic cavity			
Pleural cavities	3. ____________		
Pericardial cavity	4. ____________		
Abdominopelvic cavity			
Abdominal cavity	5. ____________	8. ____________	11. ____________
	6. ____________	9. ____________	12. ____________
	7. ____________	10. ____________	13. ____________
			14. ____________
Pelvic cavity		15. ____________	18. ____________
		16. ____________	19. ____________
		17. ____________	20. ____________
			21. ____________

D. ABDOMINOPELVIC QUADRANTS

The abdominopelvic cavity can be divided into four **quadrants** that are formed by transverse and sagittal planes running through the umbilicus (navel). These quadrants are useful clinically when trying to describe abnormalities or determining which organ may be the cause of pain. The four quadrants are: **right upper quadrant (RUQ), left upper quadrant (LUQ), right lower quadrant (RLQ),** and **left lower quadrant (LLQ).**

ACTIVITY 4 ABDOMINOPELVIC QUADRANTS

1 Draw lines on Figure 2.3 separating the abdominopelvic area into quadrants as advised by your instructor. Label the areas.

2 Using a torso model (or diagram), identify in which abdominopelvic quadrant(s) each organ is located. Use the abbreviations RUQ, LUQ, RLQ, LLQ.

a. appendix _____

b. large intestine _____

c. liver _____

d. ovaries _____

e. pancreas _____

f. small intestine _____

g. spleen _____

h. stomach _____

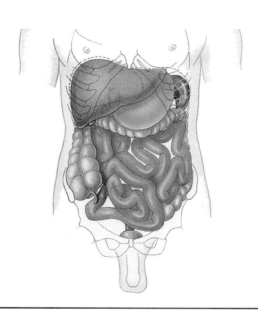

FIGURE 2.3 Abdominopelvic area.

REVIEWING YOUR KNOWLEDGE

A. Function of Organ Systems

Identify the organ system whose function is described below.

_____ **1.** Maintains blood oxygen and carbon dioxide levels

_____ **2.** Controls muscles and glands by electrical signals

_____ **3.** Controls cell function with chemical signals called hormones

_____ **4.** Movement of bones

_____ **5.** A waterproof barrier that blocks the entrance of pathogens into the body and the loss of water from the body

_____ **6.** Transports nutrients, oxygen, and carbon dioxide throughout the body

_____ **7.** Breaks down food into absorbable nutrients

_____ **8.** Regulates composition of blood by eliminating nitrogenous wastes, excess water, and minerals

_____ **9.** Protection of body organs and framework for body

_____ **10.** Produces gametes (sperm and eggs)

_____ **11.** Returns fluid that has leaked out of blood vessels into tissue, and provides protection against pathogens that have entered the body

B. Organ Identification

Identify the correct organ system for the following organs.

Organ System   **Organ**

_____ **1.** spleen

_____ **2.** liver

_____ **3.** trachea

_____ **4.** blood vessels

_____   **5.** hair

_____   **6.** kidney

_____   **7.** ovaries

_____   **8.** pituitary gland

_____   **9.** spinal cord

_____   **10.** testes

C. Body Cavities

Identify all the cavities for each organ as follows: dorsal (D), ventral (VN), cranial (C), vertebral (VR), thoracic (T), abdominal (A), or pelvic (P). All structures are present in more than one cavity.

_____   **1.** brain

_____   **2.** ovaries

_____   **3.** heart

_____   **4.** kidneys

_____   **5.** trachea

_____   **6.** liver

_____   **7.** small intestine

_____   **8.** spinal cord

_____   **9.** lungs

_____   **10.** uterus

_____   **11.** urinary bladder

_____   **12.** stomach

D. Abdominopelvic Quadrants

Name the quadrant(s) that the following organs predominantly occupy: RUQ, LUQ, RLQ, or LLQ.

_____   **1.** liver

_____   **2.** stomach

_____   **3.** spleen

_____   **4.** gallbladder

_____   **5.** appendix

_____   **6.** left kidney

_____   **7.** right ovary

THE COMPOUND LIGHT MICROSCOPE

OBJECTIVES

After completing this exercise, you should be able to:

- Describe and demonstrate how to carry, clean, use, and store the compound light microscope.
- Identify the parts of the compound light microscope and describe their function.
- Demonstrate how to place a prepared slide on the microscope, and how to properly focus a desired area of a slide under all magnifications.
- Calculate total magnification

MATERIALS

- compound light microscope
- lens paper
- prepared slides of the letter "e"
- immersion oil
- microscope slides, coverslips, flat toothpicks, and dropper bottle of dilute methylene blue

A compound light microscope is used in an anatomy and physiology class to observe small structures such as cells and tissues. The term *compound* refers to the set of lenses (ocular and one objective lens) used simultaneously to magnify the image. The term *light* refers to the necessity of using a light source to view the object.

Human cells must be magnified to be seen by the unaided human eye. The compound light microscope can magnify images up to approximately 1,000 times, depending on the magnifying power of the lenses.

A. TRANSPORTING THE MICROSCOPE

The compound light microscope is an expensive, precision instrument that must be handled appropriately. Demonstrate care in transporting, cleaning, using, and storing the microscope.

- Pick up the microscope with two hands, one holding its arm and the other supporting the base with the cord in a secure position.
- Carry the microscope upright so that a lens or eyepiece doesn't fall out and carefully place the microscope on your lab bench in front of you.

B. PARTS OF THE MICROSCOPE

1. Microscope Framework

- **Base**—The widest bottom part that supports the microscope.
- **Arm**—The straight or curved vertical part that connects the base to the upper portion.
- **Head** (or body tube)—The part that extends from the arm and contains the ocular lens and the rotating nosepiece with the objective lenses.

> **NOTE:** *All other microscope parts attach to the base, arm, and head.*

- **Ocular lenses or lens** (eyepieces)—One or two removable lenses that you look through to observe the microscope slide. Typically, these lenses have a magnification of 10×. Look at the eyepiece and record the magnification. _____ One of the oculars may have a **pointer,** used to identify a specific area on the slide, that can be used when studying slides with others.
- **Objective lenses**—Usually a microscope will have three or four objective lenses mounted on a **revolving nosepiece.** Most microscopes for anatomy and physiology classes have objective lenses with magnifications of 4× (scanning), 10× (low), 45× (high dry), and 100× (oil immersion). As the barrel of the objective lens increases in length, the magnifying power increases. List the magnifications of the objective lenses on your microscope. _____
- **Stage**—The flat platform located beneath the objective lenses on which the microscope slide is placed.
- **Mechanical stage** (or **spring clips**)—Holds the slide securely in place for viewing and can be moved with precision by using the **adjuster knobs.** If spring clips are on your microscope, adjust the slide beneath them by hand.
- **Coarse focus knobs**—On each side of the microscope toward the base is a large knob or dial with a smaller knob in the middle. The large knob is used for coarse focusing and either moves the stage up and down with the objective lens in place, or moves the objective lenses up and down with the stage in place. This knob is to be used with scanning or low-power lenses only. You need to use only one of the large knobs, depending on whether you are right- or left-handed.

- **Fine focus knob**—The smaller knob on each side of the microscope that is used for precision focusing.
- **Condenser**—Located just below the stage, this lens condenses light through the specimen on the slide above. If the condenser has an **adjustment knob** that raises and lowers the condenser, it usually needs to be in the highest position near the stage for the exercises that you will perform.
- **Iris diaphragm**—Located beneath the condenser, the iris diaphragm works similar to the iris of the eye. By adjusting its **lever,** the aperture changes diameter and regulates the amount of light that passes through the condenser. Decreasing the aperture size decreases the amount of light on the specimen and increases contrast.
- **Substage light**—The light source is usually built into the base of the microscope and has a dial on one side to control the light intensity.

2. Determining Magnification

- Total magnification is determined by multiplying your ocular lens power times the objective lens power.
- *Example:* Ocular lens power = 10×; Objective lens power = 4×; Total magnification = 40×.

ACTIVITY 1 PARTS OF THE MICROSCOPE

1 Identify the parts of your microscope as shown in Figure 3.1.

2 Compare your microscope with the one shown in Figure 3.1 and identify any differences with your lab partner.

3 Calculate the total magnification by multiplying the lens magnification powers.

_____ scanning lens × ocular lens _____ = _____ total magnification

_____ low-power lens × ocular lens _____ = _____ total magnification

_____ high-dry lens × ocular lens _____ = _____ total magnification

_____ oil immersion lens × ocular lens _____ = _____ total magnification

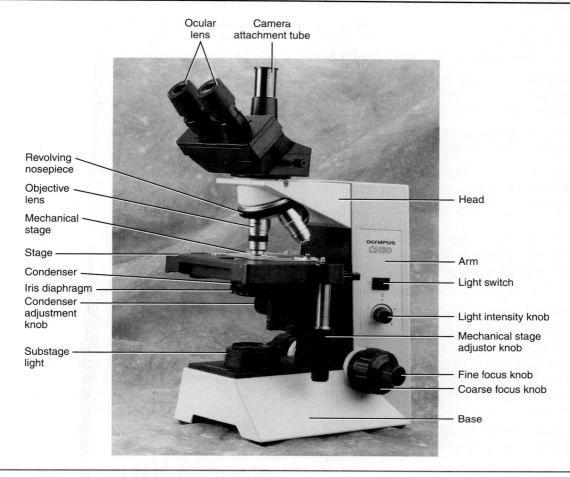

Ocular lens

Camera attachment tube

Revolving nosepiece

Objective lens

Mechanical stage

Stage

Condenser

Iris diaphragm

Condenser adjustment knob

Substage light

Head

Arm

Light switch

Light intensity knob

Mechanical stage adjustor knob

Fine focus knob

Coarse focus knob

Base

FIGURE 3.1   Parts of the microscope.

C. USING THE MICROSCOPE

- Clean up your lab area and put non-essentials away to have plenty of room to use the microscope.

- Unwind the cord and plug it in.

- Clean the ocular, objective, and condenser lenses with only the *special lens paper* (optical-safe) provided by your instructor. Whenever the image on the slide cannot be focused clearly, the objective lens may need additional cleaning. If so, consult your instructor.

- Turn on the light and adjust its intensity.

- *Troubleshooting:* If no light initially comes on, check two things before consulting your instructor. (a) Check the light dial to make sure it isn't on zero. (b) Check the safety switch on the electrical outlet by pushing in the reset button. If the light still doesn't work, plug your microscope into an outlet that you know works.

ACTIVITY 2 USING THE MICROSCOPE

1 Click the **scanning objective lens** into place so it is over the light aperture on the stage. This objective lens has the shortest barrel. Make sure you feel the click, or your view will be black.

2 Obtain a prepared slide with the letter "e" from your instructor and place it on the stage, securing it in the mechanical stage clamps or stage clips. Draw the letter "e" as it appears on the stage without looking in the microscope. _____

3 Practice moving the slide back and forth in addition to up and down using the mechanical stage knobs.

4 Using the mechanical stage knobs, position the letter "e" over the light hole in the stage and turn the light on a middle setting.

5 Check to see that the condenser lens is raised completely up to the stage.

6 If your *stage is moveable,* the coarse focus knob will move the stage. Raise the stage up as far as it will go.

If your *objective is moveable,* use the coarse focus knob to lower the objective lens until it stops. The slide and the scanning objective lens will not actually touch.

7 If you have a binocular microscope, adjust the two oculars as you would a pair of binoculars so the two oculars are a comfortable distance apart for your eyes.

8 Looking through the ocular lens(es), adjust the coarse focus knob to focus in the letter "e".

CAUTION: *Do not use the coarse focus knob with high-dry or oil immersion lenses.*

---

9 Using the mechanical stage knobs, bring the letter "e" directly into the center of the **field of view** (the lighted circular area you see as you look through the oculars).

10 Draw the letter "e" as it appears through the microscope. Compare the appearance to your initial drawing. _____

11 While observing the "e", use your mechanical stage knob to move the "e" away from you. Now move it back toward you, to your right, and to your left. What did you notice that seemed different about these movements?

12 Reposition the letter "e" directly in the middle of the field of view and switch to the **low-power objective lens.**

13 Most microscopes are **parfocal** so when you move to a different magnification the specimen is almost in focus. Use the fine focus to sharpen the image.

14 Use the iris diaphragm lever to adjust the amount of light and improve the contrast of your image.

15 Center the area you want to view and repeat this procedure with the **high-dry objective lens.** Be sure to focus *only* with the fine focus knob.

Most slides used in anatomy and physiology do not need the magnification of the oil immersion lens. Your instructor will inform you if and when you will use this objective lens.

---

16 Center the area you want to view and then obtain a container of immersion oil made especially for the oil immersion lens.

17 With the slide focused at high dry, move the objective lens out of the way and apply a drop of oil directly on the part of the slide you wish to study.

18 Click the oil immersion lens into place, open the iris diaphragm as needed, and focus *only* with the *fine focus knob.*

19 Move the scanning power objective lens in place.

20 Remove the slide and clean the oil from the objective lens with lens paper (also clean the high-dry objective lens if you passed it through the oil). Your instructor may ask you to use an additional cleaner.

21 Clean the slide with lens paper. If necessary, clean the stage as well.

22 Move the scanning power objective lens in place when done.

D. MEASURING FIELD OF VIEW

The **field of view** is the area on the slide that is being observed. The size of the field of view decreases with increasing magnification. It is important that the object being viewed is placed in the center of the field of view when switching to a higher objective lens (higher magnifying power).

ACTIVITY 3 MEASURING FIELD OF VIEW

1 Place the scanning power objective lens in place.

2 Place a clear plastic ruler over the light opening in the stage.

3 Starting with the scanning power objective, look through the ocular lens to count the number of millimeter intervals that can be seen. Record. _____ mm.

4 Switch to the low-power objective lens to count the number of millimeter intervals that can be seen. Record. _____ mm.

5 Switch to the high-dry power objective lens to count the number of millimeter intervals that can be seen. Record. _____ mm.

6 Move the scanning power objective lens in place before removing the ruler.

E. WET MOUNT OF CHEEK CELLS

Cells that line the interior of the mouth fit closely together like floor tile and form a thick cell layer that protects the underlying tissue from abrasion and microbes (bacteria and viruses). The superficial cells continually slough off and are replaced by underlying cells. Gently scraping the lining of the cheek removes the superficial cells that are about to slough off.

ACTIVITY 4 WET MOUNT OF CHEEK CELLS

1 Prepare a cheek smear slide and observe it under the compound microscope.

- Obtain a toothpick, a clean microscope slide, and a coverslip.

- Place a drop of 0.9% saline on the microscope slide.

- Gently scrape the flat end of the toothpick (no blood, please!) on the inner lining of your cheek.

- Mix the cheek cell sample in the drop of saline on the slide.

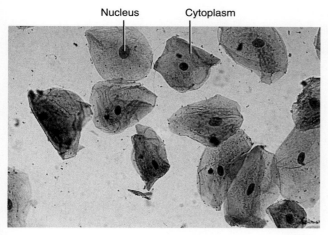

Nucleus   Cytoplasm

400×

FIGURE 3.2   Cheek cells.

- Dispose of the toothpick as your instructor directs.

- Add *one drop* of dilute methylene blue to the cells on your slide.

- Cover the sample with a coverslip as directed by your instructor.

- Using low power, locate the blue-stained cells. Switch to high power to observe more cellular detail.

2 Compare the cells on your slide with Figure 3.2.

3 Move the scanning power objective lens in place and remove the slide.

4 Place microscope slides in a 10% bleach solution as your instructor directs. Clean your lab top with a 10% bleach solution.

5 Answer the discussion question with your lab partners.

❓ DISCUSSION QUESTION: CHEEK SMEAR

1 Why was stain added to the cheek cells?

F. STORING THE MICROSCOPE

It is important to put the microscope away properly.

- Check that the scanning objective lens is in place and that the slide is removed from the stage.

- Turn off the substage light. The bulb life is extended if it cools before moving the microscope.

- Clean ocular and objective lenses with lens paper.

- Carefully wrap the cord around the base of the microscope, making sure it isn't pulled too tight.

- Place the dust cover over the microscope.

- Using both hands and the proper carrying technique, return the microscope to the appropriate cabinet.

REVIEWING YOUR KNOWLEDGE

A. Care and Use of the Microscope

Indicate what is wrong with each of these statements.

1. Only one hand is needed to transport the microscope.

2. Tissue paper can be used to clean the microscope lenses and prepared slides.

3. The microscope should be put away with the high-dry lens in position.

4. The coarse focusing knob should be used when using the high-dry lens.

5. The iris diaphragm should be completely open to obtain maximum contrast.

6. The condenser should be in the lowest position (far from stage) to focus the most light on the specimen.

B. Parts of the Microscope

Select the term that matches the descriptions.

a. arm

b. base

c. coarse adjustment knobs

d. condenser

e. fine focus knob

f. head

g. iris diaphragm

h. objective lenses

i. ocular lenses

j. stage

_____ 1. Large knob that moves the stage or objective lens a great distance. Used with scanning or low-power objective lenses only.

_____ 2. Flat platform beneath objective lens upon which the microscope slide is placed.

_____ 3. Removable lenses that you look through to observe the microscope slide.

_____ 4. Small knob that moves the stage or objective lens a very small distance and is used for precision focusing.

_____ 5. Extends from the arm and contains the ocular lenses and rotating nosepiece.

_____ 6. Lens that condenses light through the specimen and is located below the stage.

_____ 7. Light from specimen passes through these lenses first. These lenses are located in the rotating nosepiece.

_____ 8. The widest bottom part that supports the microscope.

_____ 9. Regulates the amount of light passing through the condenser.

_____ 10. The vertical portion that connects the base to the head.

C. Total Magnification and Field of View

1. Calculate the total magnification of an object viewed with a 10× ocular and a 60× objective lens.

2. Does the size of the field of view increase or decrease when going from a low to a higher power objective lens?

CELL STRUCTURE AND CELL CYCLE

OBJECTIVES

After completing this exercise, you should be able to:

- Identify cellular organelles on a model or chart
- Describe the function of the plasma membrane and cellular organelles
- Identify cells and cellular structures on prepared microscope slides
- Identify the stages of mitosis

MATERIALS

- diagram, chart, or model of a cell
- compound microscopes and lens paper
- prepared slides of skeletal muscle cells, ciliated cells (pseudostratified ciliated columnar epithelium), motor neurons, sperm, and blood
- 3-dimensional models of mitosis

A. CELLULAR STRUCTURES AND ORGANELLES

Cells are the smallest structural and functional units of living organisms. They are enclosed by a plasma membrane which controls the movement of substances into and out of the cell. The interior of the cell is filled with cytoplasm that contains cytosol (a viscous fluid) and organelles. The organelles (little organs) are structures within the cell that perform different cell functions. A "generalized" animal cell is shown in Figure 4.1, and functions of cellular organelles are described in Table 4.1.

The human body contains many different types of cells with different functions. Differences in function are reflected in cell structure. Cells differ from the generalized animal cell in shape, size, or in number and type of organelles present. In this activity you will observe skeletal muscle cells, ciliated cells, motor neurons, sperm cells, and red and white blood cells. **Skeletal muscle cells** contract (shorten in length) to move bones. **Ciliated cells** use cilia (hairlike processes) to move substances along the surface of the cells. **Motor neurons** are a type of nervous tissue cell that receive information from other neurons and send electrical signals to muscle cells causing them to contract. **Sperm cells** carry genetic information through the female reproductive tract and try to insert this genetic information into an egg (fertilization). **Red blood cells** contain large amounts of hemoglobin, a red pigment that binds oxygen. **White blood cells** defend the body from pathogens and cancerous cells. The differences in function of these cells are reflected in differences in their size, shape, and cellular organelles.

29

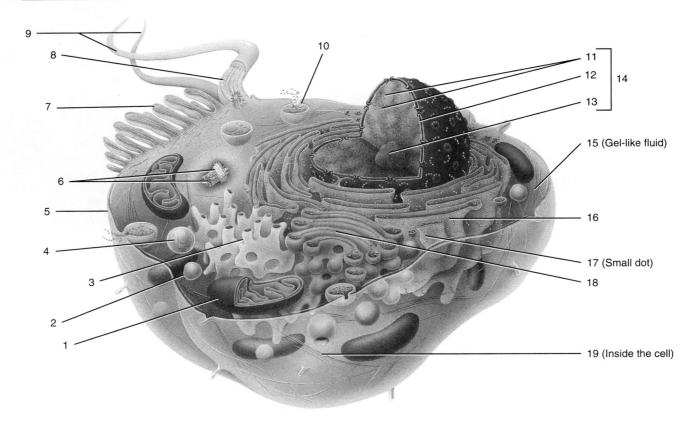

9
8
7
6
5
4
3
2
1
10
11
12
13
14
15 (Gel-like fluid)
16
17 (Small dot)
18
19 (Inside the cell)

1 mitochondrion	8 microtubules (cytoskeleton)	15 cytoplasm
2 peroxisome	9 flagella	16 rough endoplasmic reticulum
3 smooth endoplasmic reticulum	10 secretory vesicle	17 ribosome
4 lysosome	11 chromatin	18 Golgi complex
5 plasma membrane	12 nuclear membrane	19 microfilament (cytoskeleton)
6 centrioles	13 nucleolus	
7 microvillus	14 nucleus	

FIGURE 4.1   **Sectional drawing of a cell.**

TABLE 4.1 Function of Cellular Structures

STRUCTURE	FUNCTION
Plasma Membrane (cell membrane)	Mediates entrance and exit of substances into and out of the cell.
• Microvilli	Folds of the plasma membrane that increase its surface area.
Cytoplasm	Area of the cell that includes the cytosol and organelles.
Cytosol	Fluid portion of cytoplasm that surrounds organelles
Organelles	
• Nucleus (NOO-klee-us)	Contains the cell's hereditary information in genes; directs cellular activities
Nucleolus (noo-KLEE-o-lus)	Assembly site for ribosomes
Chromatin (KROH-mah-tin)	Long thin strands within nucleus. Each strand composed of one DNA molecule and associated proteins.
• Centrioles	Needed to form cilia and flagella; active in mitosis and meiosis
• Mitochondria (my-toh-KON-dree-a)	Makes ATP through aerobic cellular respiration
• Smooth endoplasmic reticulum	Phospholipid and carbohydrate synthesis site; detoxifies toxic substances
• Ribosome (RYE-bow-some)	Protein synthesis site.
• Rough endoplasmic reticulum (endo-PLAZ-mik re-TIC-u-lum)	Processes secretory proteins that are made on the ribsomes
• Golgi complex (GOAL-jee)	Membranous sacs; stores, alters, and packages cellular exports
• Lysosome (LY-so-some)	Digests worn-out organelles and substances that enter the cell
• Cytoskeleton	Three kinds of protein filaments (microfilaments, intermediate filaments, microtubules); maintains cell shape and movement
• Cilia	Move fluids and particles along a cell surface or move a one-celled organism
• Flagella	Longer than cilia; used for cell locomotion. In humans, only found on sperm cells.

ACTIVITY 1 CELL STRUCTURES AND ORGANELLES

1 Point to each cellular organelle shown in Figure 4.1 on a cell model or chart.

2 Pronounce the name of each organelle and describe its function as you point to it.

3 Using a compound microscope, observe each prepared slide of skeletal muscle, ciliated cells, motor neuron, sperm cells, and blood smear. Use Figure 4.2 to help you locate the cells.

4 Answer discussion questions with your lab partners.

❓ DISCUSSION QUESTIONS: CELL SPECIALIZATION

1 Describe each cell's shape and list the organelles that can be seen with the light microscope in each type of cell.

a. skeletal muscle cell _____

b. ciliated cells (pseudostratified ciliated columnar epithelium) _____

c. motor neuron (nerve cell) _____

d. sperm cell _____

e. red blood cell _____

f. white blood cell _____

2 What were the main differences you observed in the cells?

_____

_____

_____

Width of skeletal
muscle cell   Nucleus

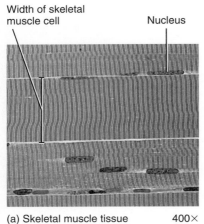

(a) Skeletal muscle tissue   400×

Nucleus   Cilia

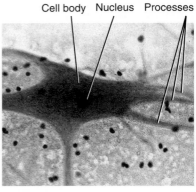

(b) Ciliated cell   400×
   (Pseudostratified ciliated columnar
   epithelium)

Cell body   Nucleus   Processes

(c) Motor neuron   350×

Nucleus   Flagellum

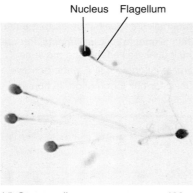

(d) Sperm cells   400×

Red blood cells   Nucleus of white
                   blood cell

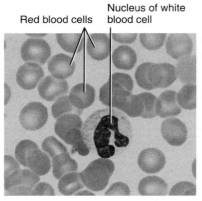

(e) Blood smear   400×

FIGURE 4.2   **Cell specialization.**

B. SOMATIC CELL DIVISION: THE CELL CYCLE AND MITOSIS

Somatic (*soma-* = body) **cell division** occurs when one cell divides to produce two identical cells. Cell division is needed for growth of the individual and cell replacement.

The cell cycle begins when a cell is produced by cell division and ends when the cell divides. The length of the cell cycle differs according to the type of cell, with some cells dividing more frequently than others. The **cell cycle** can be divided into two basic periods: **interphase,** a longer period during which the cell conducts its normal activity, grows and prepares for cell division, and the **mitotic phase,** when the cell is dividing. The mitotic phase consists of **mitosis,** or nuclear division, and **cytokinesis,** or cytoplasmic division. The four stages of mitosis are: prophase, metaphase, anaphase, and telophase (Table 4.2).

To observe interphase and the stages of mitosis, you will examine a prepared microscope slide containing several whitefish blastula. The **blastula** is an active growth stage of embryonic development that has many dividing cells.

ACTIVITY 2 MITOTIC PHASES

1 Observe the three-dimensional models of the mitotic phases, noting the changes in each phase.

2 Identify interphase, each phase of mitosis, and cytokinesis in Figure 4.3(a–e).

3 Obtain a prepared whitefish blastula slide and hold it up to the light. Notice that there are many blastulae circles placed on each slide. It will be necessary to view several of the blastulae circles to find all of the phases.

4 Using a compound microscope, begin looking at your slide with the low-power objective lens. Use the high-power objective lens to identify interphase, the stages of mitosis, and cytokinesis.

TABLE 4.2 Phases of Somatic Cell Cycle

PHASE	DESCRIPTION OF ACTIVITY
INTERPHASE (*inter-* = between)	Normal cell work; cell metabolically active and growing, DNA replicates; chromosomes not visible
MITOTIC PHASE	Parent cell divides into two daughter cells
Mitosis (*mitos-* = thread)	Nuclear division
• Prophase (*pro-* = first)	Nucleolus and nuclear membrane disappear; chromatin condenses into chromosomes; centrioles move to opposite poles; spindle fibers form
• Metaphase (*meta-* = next)	Chromosomes line up at metaphasal plate; spindle fibers attach to centromeres of chromatids
• Anaphase (*ana-* = apart)	Chromatids of chromosomes separate; move to opposite poles
• Telophase (*telo-* = end)	Cell reverses prophase activities
Cytokinesis (*cyto-* = cell; *kinesio-* = movement)	Cytoplasmic division into two daughter cells

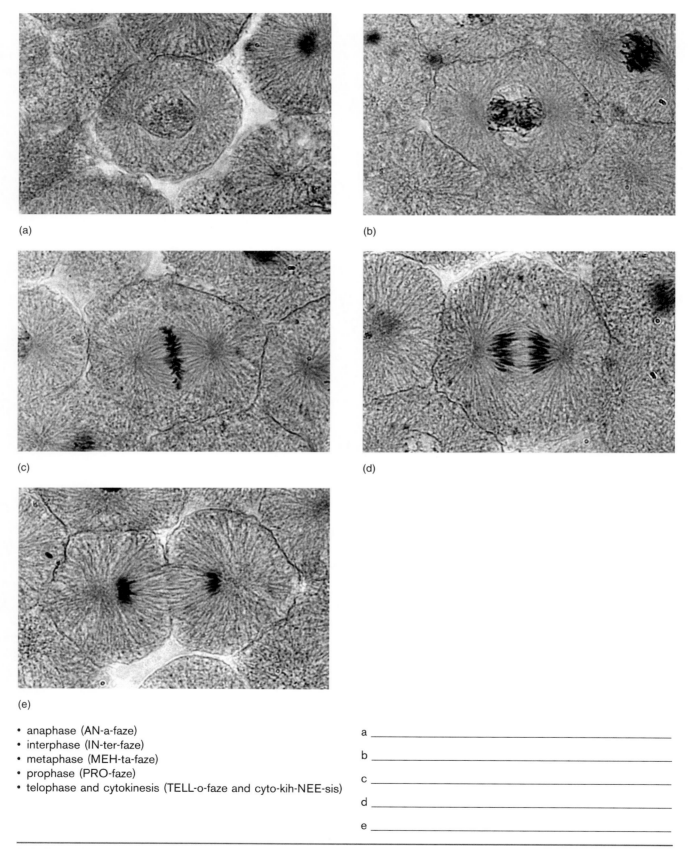

(a)

(b)

(c)

(d)

(e)

- anaphase (AN-a-faze)
- interphase (IN-ter-faze)
- metaphase (MEH-ta-faze)
- prophase (PRO-faze)
- telophase and cytokinesis (TELL-o-faze and cyto-kih-NEE-sis)

a _____

b _____

c _____

d _____

e _____

FIGURE 4.3 **Mitotic phases.**

REVIEWING YOUR KNOWLEDGE

A. Cellular Structures and Organelles

Fill in the blank with the name of the cell structure or organelle that fits the description.

_____ 1. short hair-like projections for movement of substances over cell surface

_____ 2. intracellular fluid

_____ 3. ATP production by cellular respiration

_____ 4. protein synthesis site

_____ 5. carbohydrate and lipid synthesis site

_____ 6. small vesicle with digestive enzymes

_____ 7. organelles needed to form cilia and flagella

_____ 8. thread-like strand of DNA with associated proteins

_____ 9. synthesis of secretory proteins

_____ 10. stores, packages, and exports protein product

_____ 11. contains genes that control cellular activities

_____ 12. site for synthesis of ribosomes

_____ 13. maintains cell shape and cell movement

_____ 14. mediates transport of substances into or out of the cell

_____ 15. long, hair-like projection that moves the cell

_____ 16. folds of plasma membrane that increase the surface area of cell

B. Phases of the Cell Cycle

Match the description with the phase of the cell cycle.

_____ **1.** anaphase

_____ **2.** cytokinesis

_____ **3.** interphase

_____ **4.** metaphase

_____ **5.** mitosis

_____ **6.** prophase

_____ **7.** telophase

a. Cytoplasmic division

b. Cell performing normal functions; longest phase

c. Nuclear division

d. Chromatid pairs line up at equatorial plate

e. Chromatin condenses into chromosomes; nucleus disappears

f. Spindle fibers break up; nucleus reappears; chromosomes unravel to form chromatin

g. Centromeres divide; chromosomes move to opposite poles

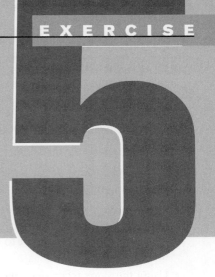

TRANSPORT ACROSS CELL MEMBRANES

OBJECTIVES

After completing this exercise, you should be able to:

- Describe diffusion, osmosis, and filtration
- Distinguish between and define hypotonic, hypertonic, and isotonic solutions
- State the direction of net water movement when a cell is placed in a hypotonic, hypertonic or isotonic solution

MATERIALS

- Diffusion: 1 agar petri dish per group, small millimeter rulers, forceps, methylene blue crystals, potassium permanganate crystals
- Osmosis in living cells: animal blood from biological supply house, disposable gloves, clean microscope slides, coverslips, 4 medicine droppers per group, compound microscope, filter paper, distilled water, 0.9% saline solution, 5% saline solution, and 10% bleach solution (clean up)
- Filtration: filter paper, funnel, ring stand with ring, 250-mL beaker, distilled water, 5% powdered charcoal solution, and 1% copper sulfate solution

The plasma membrane (cell membrane) with its unique design is responsible for selectively allowing substances into and out of the cell either by active or passive transport processes. **Active transport processes** require ATP (stored energy) to move substances across the plasma membrane. **Passive transport processes** do not require the cell to expend energy because the kinetic energy of the particles causes them to move from an area of their higher concentration to an area of their lower concentration. In this exercise you will observe the following passive transport processes: diffusion, osmosis, and filtration.

A. DIFFUSION

In **diffusion,** there is a net movement of substances from a region of their greater concentration to a region of their lesser concentration. Diffusion can occur in a solid, liquid, or gas, and across the plasma membrane of cells. In the following activity, diffusion of two crystals through a solid (agar) will be studied. Methylene blue has a molecular weight of 320, and potassium permanganate has a molecular weight of 158. The two crystals move at different rates through the agar, which is composed primarily of water.

ACTIVITY 1 DIFFUSION

1 Using the molecular weight of the two crystals, predict which crystal will move faster. Circle your answer: *methylene blue* or *potassium permanganate.*

2 Obtain materials for this activity.

3 Measure the diffusion rates of methylene blue and potassium permanganate.

   • Using forceps, carefully place a large crystal of methylene blue on the surface of an agar petri dish (Figure 5.1). Be careful not to drop any extra crystals on the agar surface.

   • Using the same technique, place a similar sized crystal of potassium permanganate on the other side of the petri dish.

   • Using a millimeter (mm) ruler, measure the crystals' diameter of diffusion at 15-minute intervals for at least 1 hour (or longer if desired).

   • After each observation, record the diffusion diameter of each crystal in millimeters (mm) in Table 5.1.

4 Answer the discussion questions with your lab partners.

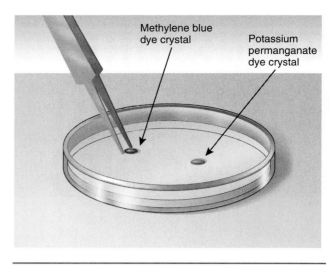

FIGURE 5.1 **Diffusion in agar plate setup.**

❓ DISCUSSION QUESTIONS: DIFFUSION

1 Which substance moved faster?

2 Why did that substance move faster?

TABLE 5.1 Diffusion Results

TIME (MIN)	DIFFUSION OF METHYLENE BLUE (DIFFUSION DIAMETER IN MM)	DIFFUSION OF POTASSIUM PERMANGANATE (DIFFUSION DIAMETER IN MM)
15		
30		
45		
60		
75		

B. OSMOSIS

Osmosis is the net *diffusion* of *water molecules* through a selectively permeable membrane from a solution with a higher water concentration and a lower solute concentration (**hypotonic**) into a solution with a lower water concentration and a higher solute concentration (**hypertonic**). **Solutes** are the substances dissolved in a solvent. A **solvent** is a dissolving medium. Water and lipids are solvents in humans, with water being the main solvent. The terms hypotonic (*hypo-* = deficient; *-tonos* = stretching) solutions or hypertonic (*hyper-* = excessive) solutions are used only when two solutions are compared. Water will move into the hypertonic solution until the water and solute concentrations of the two solutions equalize to become **isotonic** solutions (*iso-* = same).

Red blood cells (RBCs) are good examples to use for an osmosis experiment because their shape changes dramatically when they are exposed to hypotonic or hypertonic solutions. Their normal shape is a round, biconcave disk with a smooth plasma membrane. If the cell loses most of its water by osmosis when put in a hypertonic solution, it becomes **crenated** or shriveled with spiked edges. If the cell gains a significant amount of water by being placed in a hypotonic solution, it swells and may eventually burst—a process called **hemolysis** (*hemo-* = blood; *-lysis* = break down).

The salt content of intracellular and extracellular fluids in the body is 0.9%. Physiologic saline contains 0.9% of the salt NaCl.

> **NOTE:** *Remember . . . salt shrivels and water swells a cell.*

Safety Precautions

- Clean laboratory tabletops before and after conducting your experiments.
- Wear safety glasses and disposable gloves when handling body fluids or animal blood.

ACTIVITY 2 OSMOSIS IN LIVING CELLS

1 With your lab group, discuss and identify which solutions for this experiment are hypotonic, hypertonic, or isotonic to the RBCs.
- 0.9% saline solution (physiologic saline) _____
- 5% saline solution _____
- distilled water _____

2 With your lab partners, predict the net movement of water in all three types of solutions. Circle your choices for all three situations:

- In the isotonic solution, the RBCs will *swell*, *crenate*, or *not change shape*.
- In the hypotonic solution, the RBCs will *swell*, *crenate*, or *not change shape*.
- In the hypertonic solution, the RBCs will *swell*, *crenate*, or *not change shape*.

3 In Figure 5.2, identify the shape of the RBC (swollen, crenated, or normal) and the type of solution the RBC is in (isotonic, hypertonic, or hypotonic).

4 Observe osmosis in RBCs.

Setup
- Obtain materials for the activity
- Take out a compound microscope
- Put on gloves

Experimental Procedure
Slide #1: Control
- Place a drop of animal blood on a slide and cover with a coverslip. Observe the RBCs using high power. Blood cells are very tiny and difficult to see under low power. Record the shape of the cell _____.

Slide #1: 5% saline solution
- Using slide 1 again, add one drop of 5% saline to one side of the coverslip. Place filter paper or a small piece of paper towel on the opposite edge of the coverslip to absorb the liquid and pull the 5% saline solution into the RBCs.
- Immediately observe the slide under high power and watch for changes in cell shape. You may have to wait a few minutes for cell changes to occur.
- Record the shape of the RBCs and type of solution in which they were placed.
shape _____ type of solution _____

Slide #1: distilled water
- Using slide #1 again, add one drop of distilled water to the same edge of the coverslip that was used for the saline solution.
- Hold a piece of filter paper at the opposite edge of the coverslip, observe the paper absorb the saline solution, and watch as the distilled water is drawn under the coverslip and into the RBCs.
- Immediately observe the slide under high power and watch if the RBCs change shape.
- Record the shape of the RBCs and type of solution in which they were placed.
shape _____ type of solution _____

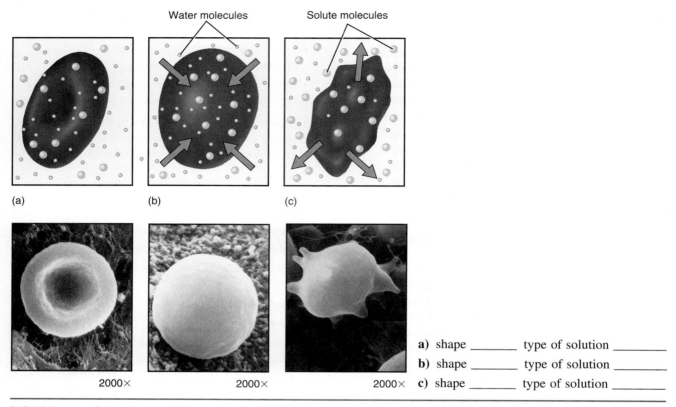

a) shape _____ type of solution _____
b) shape _____ type of solution _____
c) shape _____ type of solution _____

FIGURE 5.2 Osmosis in RBCs.

Slide #2: 0.9% saline solution

- Place a drop of animal blood on a clean slide with a medicine dropper. With a different medicine dropper, add a drop of 0.9% saline solution to the blood. Tilt the slide to intermix the two solutions, and cover with a coverslip.

- Using high power, observe the slide for changes in cell shape.

- Record the shape of the red blood cells and the type of solution in which they were placed.

  shape _____ type of solution _____

Clean Up

- Place blood-stained items (slides, coverslips, and droppers) in a 10% bleach solution or as directed by your instructor.

- Place gloves in an autoclavable bag or location indicated by your instructor.

- Wash down the lab counters with 10% bleach solution in a squirt bottle and wipe with paper towels wet with 10% bleach solution. Allow to air dry.

- Wash your hands with soap and water before leaving the lab area.

5 Answer discussion questions and conclusions with your lab partners.

❓ DISCUSSION QUESTIONS: OSMOSIS IN RBCs

1 Does distilled water contain any solutes?

2 Does tap water contain any solutes?

3 Compare what would happen if the RBCs were placed in tap water instead of distilled water.

4 Osmosis was observed using which of the three solutions in Activity 2?

3 Set up and conduct the experiment (Figure 5.3).

- Decide who will be the timer, the counter, and the recorder in your group.

- Fold the circular filter paper in half twice, and open it to form a cone. Place the cone in the funnel and place the funnel in a ring on the ring stand. Put the beaker under the funnel.

- Stir or mix the filtrate solution comprised of water, copper sulfate (blue), and powdered charcoal, and pour it into the funnel to just below the top of the filter paper. Watch so the edges of the filter paper do not capsize or that the solution doesn't go over the top.

- As soon as the liquid slows down enough to count individual drops, count the number of drops in 15-second intervals.

- Record your information on Table 5.2.

- Observe the filtrate (liquid) in the beaker. If it is blue, the copper sulfate went through the filter; if it is black, the charcoal went through the filter.

- Compare your results with your predictions.

4 Answer the discussion questions with your lab partners.

C. FILTRATION

Filtration is the separation of substances passed through a filter or membrane from a region of higher pressure to a region of lower pressure. In the human body, filtration occurs in the kidneys and capillaries. Liquid and dissolved solutes flow through the pores of a filter or membrane, keeping solutes larger than the pores on the original side. In this activity, you will observe **selectivity** of the filtration membrane and a **change in filtration rate.** Keep in mind that the plasma membrane is part of a living cell and can have active processes going on simultaneously with passive processes such as filtration.

ACTIVITY 3 FILTRATION

1 Obtain materials for this activity.

2 With your lab partners, predict which substances will pass through the filter: water, copper sulfate (blue), or charcoal.

_____ will pass through the filter.

_____ will not pass through the filter.

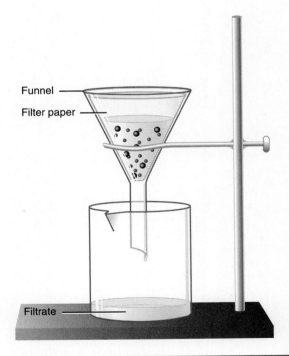

Funnel

Filter paper

Filtrate

FIGURE 5.3   Filtration setup.

TABLE 5.2 Filtration Results

TIME (SEC)	#DROPS
0–15	
16–30	
31–45	
46–60	
61–75	
76–90	

? DISCUSSION QUESTIONS: FILTRATION

1 What cell structure does the filter paper represent?

2 What is the force that drives the filtration?

3 Did you see a difference in the filtration rate during the activity? Explain.

REVIEWING YOUR KNOWLEDGE

A. Transport Across the Plasma Membrane

Match the definition with the term.

_____ **1.** solvent

_____ **2.** crenate

_____ **3.** diffusion

_____ **4.** hemolysis

_____ **5.** hypertonic solution

_____ **6.** hypotonic solution

_____ **7.** isotonic solution

_____ **8.** osmosis

_____ **9.** solute

a. to shrink or shrivel

b. water moving through selectively permeable membrane

c. random mixing of particles from their greater concentration to their lesser concentration

d. substance dissolved in a solution

e. to burst a red blood cell

f. same solute concentration on both sides of plasma (cell) membrane

g. solution with a lower concentration of solutes than in the cytosol of the cell

h. a fluid that can contain dissolved substances

i. solution with a higher concentration of solutes than in the cytosol of the cell

B. Osmosis

1. A test tube with RBCs in it has a solution added to it. After several minutes, the solution turns red. What happened to the RBCs?

2. Which solution was added to the RBCs to obtain these results?

A. 0.9% saline   B. 5% saline   C. distilled water

3. An 0.8% saline solution would be _____ to the cytosol of a cell.

A. hypotonic   B. hypertonic   C. isotonic

4. If you placed a peeled apple or potato in a 5% salt solution, it would:

A. gain weight   B. lose weight   C. stay the same weight

5. A person's hands become wrinkled after spending a long, relaxing time in a bathtub. Tub water does not have as many solutes in it compared with the human body. Explain why the hands look wrinkled.

A. The tub water is *hypotonic* to body cells and water *enters* the cells.

B. The tub water is *hypotonic* to body cells and water *leaves* the cells.

C. The tub water is *hypertonic* to body cells and water *enters* the cells.

D. The tub water is *hypertonic* to body cells and water *leaves* the cells.

TISSUES

OBJECTIVES

After completing this exercise, you should be able to:

- Identify the four primary tissue types
- Compare and contrast primary tissue structure and function
- Identify examples of epithelial, connective, muscular, and nervous tissue types and describe their location and function

MATERIALS

- compound microscope and lens paper
- prepared epithelial tissue slides: simple squamous (tissue sections of mesothelium), simple cuboidal (kidney), simple columnar nonciliated (small intestine), stratified squamous nonkeratinized (esophagus), transitional (relaxed urinary bladder), and ciliated pseudostratified columnar (trachea)
- prepared connective tissue slides: areolar, reticular, adipose, dense regular (tendon), dense irregular (skin), hyaline cartilage (trachea), bone (dried compact bone), and blood
- prepared muscle tissue slides: skeletal muscle, cardiac muscle, and smooth muscle
- prepared motor neuron slide
- colored pencils

There are four primary tissues: epithelial, connective, muscle, and nervous. **Epithelial tissue** covers surfaces, lines cavities, and forms glands. **Connective tissue** supports and protects other tissues. **Muscle tissue** causes movement. **Nervous tissue** receives and generates nerve impulses. Organs contain two or more tissues working together to perform a specific function.

A. EPITHELIAL TISSUE

There are two categories of epithelial tissue—covering and lining epithelium and glandular epithelium. **Covering and lining epithelium** covers the surface of the body and some organs, and lines all hollow body structures. This epithelium forms barriers that protect and control what substances can cross into adjacent tissues. **Glandular epithelium** forms glands that produce and secrete products needed by the body. **Exocrine** (*exo-* = outside; *-crine* = secrete) glands secrete their product through a duct onto a

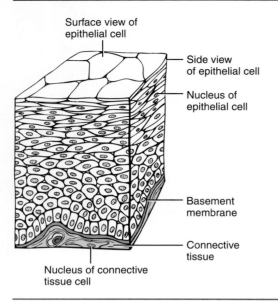

Surface view of
epithelial cell

Side view
of epithelial cell

Nucleus of
epithelial cell

Basement
membrane

Connective
tissue

Nucleus of connective
tissue cell

FIGURE 6.1 **Epithelial tissue lining the mouth.**

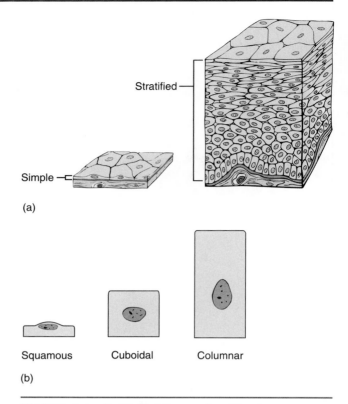

Stratified

Simple

(a)

Squamous   Cuboidal   Columnar

(b)

FIGURE 6.2 **Epithelial tissue classification.**

body surface such as the skin or lumen of a hollow organ. **Endocrine** (*endo-* = within) glands are ductless glands that produce and secrete hormones that diffuse into blood vessels.

Epithelial tissues exhibit **cellularity** (most of the tissue consists of cells and not extracellular material) and are supported by underlying connective tissue (Figure 6.1). Epithelial and connective tissues are separated by a **basement membrane.** Epithelial cells are packed together in an orderly fashion. Cell membranes of adjacent cells may be fused, forming tight junctions.

Covering and lining epithelium faces a space such as the **lumen** (*lumen* = light) of a blood vessel or intestine. The epithelial cell surface adjacent to the space is called the **apical** (*apex* = tip) **surface,** whereas the epithelial cell surface adjacent to the basement membrane is the **basal surface.** When viewed from the apical surface, one can observe how epithelial cells fit together like floor tiles to form a lining or barrier (Figure 6.1). When viewed in cross-section (side view of cells), one can observe differences in structure of the apical and basal surfaces. Apical surfaces of epithelial tissues may contain microvilli or cilia. Microvilli increase the surface area for absorption, while cilia move substances over the cell surface.

1. Epithelial Tissue Types

Epithelial tissue types are classified according to the number of epithelial cell layers (Figure 6.2) and the shape of the cell in the apical cell layer. If there is only one cell layer, the epithelium is **simple epithelium;** if there are two or more cell layers, the epithelium is **stratified epithelium.**

Epithelial cells have four shapes: squamous, cuboidal, columnar, and transitional. Cell shapes are best seen in side

views of the cells (Figure 6.2). **Squamous cells** have a flattened nucleus and are the thinnest cells. **Cuboidal cells** are cube-like with a round nucleus in the center of the cell. **Columnar cells** are tall with an oval nucleus close to the base of the cell. **Transitional cells** change shape; they are rounder when the tissue is relaxed and flatten when the tissue is stretched.

Simple epithelium provides a selective barrier allowing diffusion, filtration, secretion, or absorption of selected substances. There are three types of simple epithelium: simple squamous, simple cuboidal, and simple columnar. **Stratified epithelium** is thicker, is subject to wear and tear, and forms a protective barrier. Multiple cell layers make the tissue more resistant to damage, thereby preventing pathogens and foreign materials from crossing into underlying tissues. There are four types of stratified epithelium: stratified squamous, stratified cuboidal, stratified columnar, and transitional. Another type of epithelium is **pseudostratified** (*pseudo-* = false) **columnar epithelium.** This epithelium is only one cell layer thick because all the cells touch the basement membrane, although not all cells reach the apical surface. Therefore, there are cells of different shapes and heights, and their nuclei are at different levels. This gives the illusion of several different layers of cells. The tallest cells are narrow where they touch the basement membrane but have a columnar shape toward the outer surface, while the shorter cells do not reach the apical surface.

NOTE: *Observe tissue first with a scanning or low-power objective lens, focusing with the coarse focus knob. Before switching to a more powerful objective lens, move the slide so that the area you want to observe is in the center of the field of view. After changing to a more powerful lens, use the fine focus knob only to bring the tissue into focus. Use this procedure with all tissue slides.*

ACTIVITY 1 MICROSCOPIC EXAMINATION OF EPITHELIAL TISSUE TYPES

1 Examine and color the stratified epithelium in Figure 6.1. Color the apical and basal layers of the stratified squamous epithelium with contrasting colors.

2 Examine the photomicrographs of the epithelial tissue types in Figures 6.3–6.8. Stratified cuboidal and stratified columnar are not included because they are less common.

a. Survey photomicrograph (40×) for each tissue

• Observe that the epithelial and connective tissue layers stain differently.

• Locate the basement membrane that forms the border between the epithelial and connective tissue layers.

• Note whether there is one epithelial cell layer (simple epithelium) or two or more epithelial cell layers (stratified epithelium).

b. Photomicrograph (400×) for each tissue

• Label the photomicrographs.

• Examine the shape of the epithelial cell at the apical surface.

• Note any structural differences between apical surface and basal surface of epithelial tissue.

3 Review the location and function for each epithelial tissue type in Tables 6.1–6.6.

4 Examine prepared microscope slides of the different epithelial tissue types. Use the survey photomicrographs to help you locate the epithelial tissue at low power. Use the photomicrographs at 400× to help you locate the major structures in each tissue.

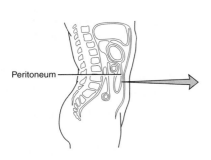

Peritoneum

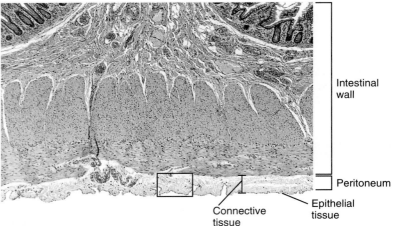

Intestinal wall

Peritoneum

Connective tissue

Epithelial tissue

(a) Survey photomicrograph, 40×

- connective tissue
- nucleus of connective tissue cell
- nucleus of simple squamous epithelial cell
- simple squamous epithelium

1 _____

2 _____

3 _____

4 _____

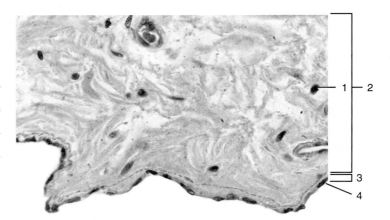

1 2

3

4

(b) Simple squamous epithelium, 400×, H&E, human

FIGURE 6.3 Sectional view of peritoneum.

TABLE 6.1 Location and Function of Selected Simple Squamous Epithelium

LOCATION	FUNCTION
Mesothelium (epithelial layer of peritoneum)	**Secretion** of serous fluid into peritoneal cavity
Alveoli (air sacs of lungs)	Short distance for **diffusion** of oxygen and carbon dioxide
Glomerular capsule (part of filtration membrane in kidney)	**Filtration** of blood to form urine filtrate (substance that is converted into urine)
Endothelium of capillaries	Short distance for **diffusion** of substances between blood and interstitial fluid (tissue fluid)

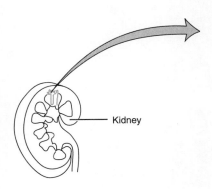

Kidney

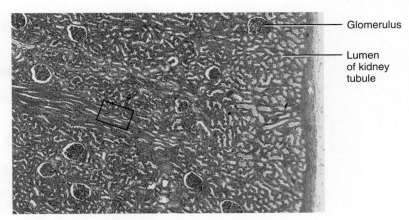

Glomerulus

Lumen
of kidney
tubule

(a) Survey photomicrograph, 40×

- apical surface of simple cuboidal epithelial
  cell
- lumen of kidney tubule
- nucleus of simple cuboidal epithelial cell
- simple cuboidal epithelium

1 _____

2 _____

3 _____

4 _____

Question: Explain why some tubule cross-
sections are circular and others are oval.

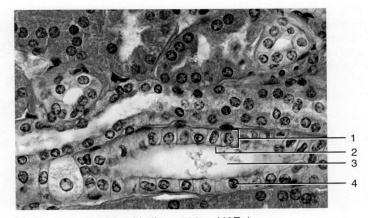

1
2
3

4

(b) Simple cuboidal epithelium, 400×, H&E, human

FIGURE 6.4   Sectional view of kidney tubules.

TABLE 6.2 Location and Function of Selected Simple Cuboidal Epithelium

LOCATION	FUNCTION
Walls of kidney tubules	Modify urine filtrate by **absorption** of substances from the filtrate and **secretion** of other substances into the filtrate
Glands	**Secretion** of products made by the simple cuboidal epithelial cells

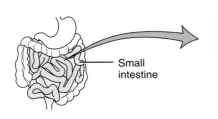

Small intestine

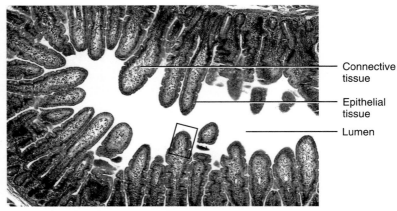

— Connective tissue

— Epithelial tissue

— Lumen

(a) Survey photomicrograph, 40×

- connective tissue
- microvilli on apical surface of simple columnar epithelial cell
- nucleus of simple columnar epithelial cell
- simple columnar epithelium

1 _____

2 _____

3 _____

4 _____

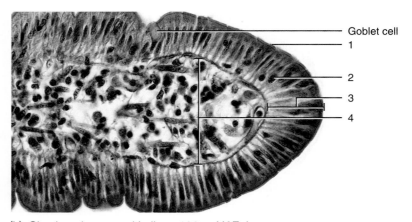

— Goblet cell
— 1
— 2
— 3
— 4

(b) Simple columnar epithelium, 400×, H&E, human

FIGURE 6.5 **Sectional view of small intestine.**

TABLE 6.3	Location and Function of Selected Simple Columnar Epithelium
LOCATION	**FUNCTION**
Lining of stomach and intestines	**Secretion** of digestive juices by simple columnar cells; secretion of mucus by goblet cells. In the small intestine the simple columnar epithelial cells have **microvilli** (*micro-* = small; *-villi* = shaggy, hair-like) to increase surface area for **absorption**
Uterine tubes (Fallopian tubes)	Simple columnar epithelial cells have **cilia** that help **move** the egg to the uterus
Central canal of spinal cord	Ciliated cells **move** cerebrospinal fluid

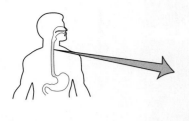

Mucous
glands

Lumen

Epithelial
tissue

Connective
tissue

Smooth
muscle

(a) Survey photomicrograph, 40×

- connective tissue
- nucleus of epithelial cell in basal layer of epithelium
- nucleus of squamous epithelial cell
- stratified squamous epithelium

1 _____

2 _____

3 _____

4 _____

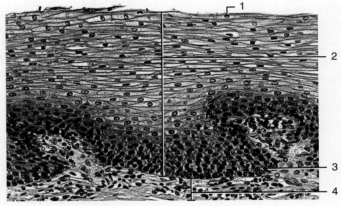

(b) Nonkeratinized stratified squamous epithelium, 200×, H&E, human

FIGURE 6.6   Sectional view of esophagus.

TABLE 6.4 Location and Function of Selected Stratified Squamous Epithelium

LOCATION	FUNCTION
Surface of skin (keratinized stratified squamous epithelium)	Epithelial layer of skin is a tough, dry, waterproof outer surface that forms a **protective barrier**
Lining of mouth, esophagus, anus, and vagina (nonkeratinized stratified squamous epithelium)	Moist epithelial layer that forms a **protective barrier** in areas subject to abrasion and friction

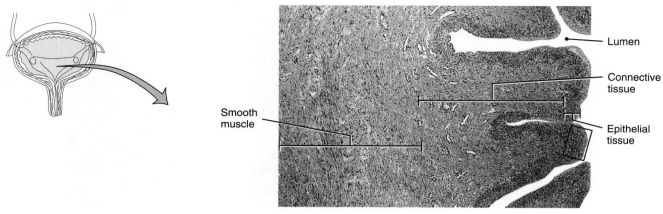

(a) Survey photomicrograph, 40×

- connective tissue
- nucleus of transitional epithelial cell in apical layer
- nucleus of transitional epithelial cell in basal layer
- transitional epithelium

1 _____

2 _____

3 _____

4 _____

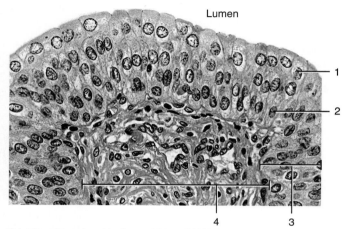

(b) Transitional epithelium, 400×, H&E, human

FIGURE 6.7 **Sectional view of a relaxed urinary bladder.**

TABLE 6.5 Location and Function of Transitional Epithelium	
LOCATION	**FUNCTION**
Lines urinary bladder and parts of the ureters and the urethra	Provides a **protective barrier** that permits **distension**

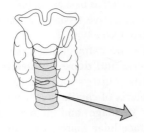

Lumen

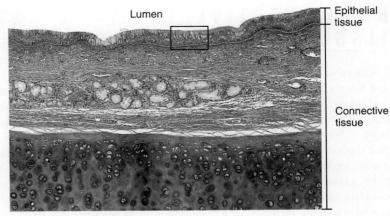

Epithelial tissue

Connective tissue

(a) Survey photomicrograph, 40×

- cilia on apical surface of columnar epithelial cell
- connective tissue
- nucleus of ciliated columnar epithelial cell
- pseudostratified ciliated columnar epithelium

1 _____

2 _____

3 _____

4 _____

goblet cell   1

2

3

4

(b) Pseudostratified ciliated columnar epithelium, 400×, H&E, human

FIGURE 6.8 **Sectional view of the trachea.**

TABLE 6.6 **Location and Function of Selected Pseudostratified Columnar Epithelium**

LOCATION	FUNCTION
Lines nasal cavity, trachea, and bronchi	**Secretion** of mucus by goblet cells. The columnar cells have **cilia** which **move** the mucus toward the pharynx.

B. CONNECTIVE TISSUE OVERVIEW

Connective tissue is the most abundant tissue in the body and has a variety of functions. It supports epithelial tissue and connects it to other tissues. Connective tissue forms tendons (connects muscle to bone) and ligaments (connects bone to bone). Connective tissue also provides coverings that support and protect muscle and nervous tissue. Bone, the hardest connective tissue, protects body organs and provides a framework for movement of muscles. Adipose tissue (fat) insulates body tissues and stores lipids, and blood provides a liquid medium for transportation of substances throughout the body.

Most connective tissues are not very cellular, usually containing more extracellular material (matrix) than cells (Figure 6.9). **Extracellular matrix** consists of **ground substance** and **fibers** that are synthesized and secreted by connective tissue cells.

Connective tissue function is determined by the properties of the extracellular matrix components. The three major types of fibers are **collagen fibers** (provide strength), **elastin fibers** (provide strength and elasticity), and **reticular fibers** (fine, branching collagen fibers that provide a net-like framework). The number and type of fibers present contribute to the strength, elasticity, and structure of the extracellular matrix. The **ground substance** may be fluid, semifluid, gelatinous, or hard. It secures the connective tissue cells and fibers in the extracellular matrix and is the medium through which interstitial fluid diffuses between blood and cells and through which macrophages and white blood cells move.

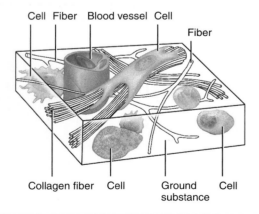

FIGURE 6.9 **Connective tissue (areolar).**

Mature connective tissues include loose connective tissue, dense connective tissue, cartilage, bone, and blood. The connective tissue types have different connective tissue cells forming the extracellular matrix. The extracellular matrix of **loose and dense connective tissues** is formed by **fibroblasts** (*blast* = early, developing stage); the extracellular matrix of **cartilage** is formed by **chondroblasts** (*chondro-* = cartilage) and maintained by **chondrocytes;** and the extracellular matrix of **bone** is formed by **osteoblasts** (*osteo-* = bone) and maintained by **osteocytes.** **Blood** was considered a connective tissue by early histologists because it connected tissues by transporting substances from one to another. **Plasma** is the extracellular matrix in which blood cells and platelets are suspended; however, it is not made by these cells.

Connective Tissue Types

1. Loose Connective Tissue

Loose connective tissue includes areolar, reticular, and adipose tissues. In **loose connective tissue,** the fibers in the extracellular matrix are loosely arranged. Collagen, elastic, and reticular fibers in this tissue provide strength, elasticity, and support. The ground substance is semifluid (viscous), but interstitial fluid can easily diffuse through it. Fibroblasts and adipocytes (lipid-storing cells) permanently reside in the tissue. Other cells, such as macrophages, mast cells, and white blood cells, enter the connective tissue from blood and are involved in body defenses.

Areolar connective tissue is the most abundant connective tissue. It contains all three fiber types, fibroblasts, a variety of cells involved in body defenses, and a viscous ground substance. **Reticular** (*reticulo-* = net-like) **fibers** are the dominant fiber type in **reticular connective tissue.** Reticular cells or **reticulocytes** synthesize the reticular fibers and ground substance. Reticulocytes are actually fibroblasts, but they synthesize more reticular fibers than collagen fibers. **Adipose** (*adipo-* = pertaining to fat) **tissue,** like areolar connective tissue, contains fibroblasts, fibers, ground substance, and adipose cells. However, adipose tissue has a greater number of adipose cells or **adipocytes** that store lipid. The lipid within the cell occupies most of the cell volume and pushes cytoplasm and organelles to the periphery. Some slides do not have the lipid stained, or the lipid has been removed during tissue preparation. In those slides, the nucleus and cytoplasm are stained, and the space inside the cell normally occupied by lipid appears "empty."

2. Dense Connective Tissue

The extracellular matrix of dense connective tissue is packed with fibers and contains very little ground substance and few fibroblasts. The extracellular matrix of **dense regular connective tissue** is packed with parallel bundles of collagen fibers that run in the direction of the pulling forces applied to the tissue. Fibroblasts are squeezed between the bundles, and the cytoplasm extends between and around collagen bundles. Fibroblasts secrete collagen fibers and ground substance. Collagen fibers give unstained tissue a silvery white appearance, which is why this tissue is often called white fibrous tissue.

Dense irregular connective tissue, like dense regular connective tissue, has little ground substance and few fibroblasts. The extracellular space is also packed with bundles of collagen fibers with fibroblasts squeezed between these bundles. In dense irregular connective tissue, however, the bundles of collagen fibers are irregularly arranged. This reflects the direction of the pulling forces to which this tissue is exposed, usually from many different directions. In **elastic connective tissue,** the extracellular matrix is packed with elastic fibers, and fibroblasts are found in the spaces between these fibers. Elastic fibers allow the tissue to be stretched and then regain its original size and shape (recoil).

3. Cartilage

Cartilage is a specialized form of connective tissue. Its extracellular matrix consists of collagen and elastic fibers embedded in a gelatinous ground substance. Collagen fibers give the tissue its strength, and the ground substance, which binds water to form a gel, gives cartilage resiliency. **Chondroblasts** (*chondro-* = cartilage) secrete the fibers and ground substance, become isolated in spaces called **lacunae** (little lakes), and transform into **chondrocytes.** Since the tissue is avascular, the chondrocytes receive their nutrients through diffusion from adjacent vascular tissues.

There are three types of cartilage: hyaline (*hyalos* = glass), elastic, and fibrous. These cartilage types differ in the amount and kinds of fibers and ground substance molecules present in their extracellular matrix. Hyaline cartilage is the most common cartilage in the body and contains fine collagen fibers in the extracelluar matrix, while elastic cartilage contains elastic fibers in the extracellular matrix. These fibers provide elasticity. The extracellular matrix of fibrocartilage contains bundles of collagen fibers which provide strength and flexibility.

4. Bone

Bone is the hardest of the connective tissues. It protects body organs and provides support for organs and skeletal muscles. The extracellular matrix is organized in layers called **lamellae** and consists of collagen fibers, ground substance, and inorganic salts. Inorganic salts, especially calcium salts, give bone its hardness; collagen fibers provide strength and flexibility. **Osteocytes** (mature bone cells) are trapped in spaces called **lacunae.** Small canals called **canaliculi** are two-way canals that connect lacunae to larger canals that contain blood vessels. Nutrients diffuse from a blood vessel through the canaliculi to the osteocytes, and waste materials diffuse back to the blood vessel for removal.

The two types of bone tissue are **compact bone** and **spongy bone.** These two types of osseous tissue differ in the amount and size of spaces present. The extracellular matrix of compact bone is organized into repeating structural units called **osteons.** In the center of each osteon is a large central canal. Spongy bone has large spaces compared to compact bone and does not have osteons.

5. Blood

Blood is considered a connective tissue because it provides a transport system that connects different body tissues. Nutrients and chemical messengers are transported to all body tissues by blood, and wastes are transported from tissues to excretory organs through the blood. Blood is composed of **red blood cells, white blood cells, platelets,** and plasma. **Plasma** is the extracellular matrix with fibers that are observed only during blood clotting. Red blood cells obtain their red color from hemoglobin, and the mature cells do not contain a nucleus (anucleate). White blood cells, which appear white or clear on unstained slides, are nucleated. Platelets, anucleate fragments of cells that are much smaller than red blood cells, are active in blood clotting.

ACTIVITY 2 MICROSCOPIC EXAMINATION OF CONNECTIVE TISSUE TYPES

1 Examine and color the areolar connective tissue in Figure 6.9. Color the different fiber types with contrasting colors.

2 Label and examine the photomicrographs of the connective tissue types in Figures 6.10–6.18.
  - Observe the extracellular matrix in each tissue.
  - Compare the amount of fibers and ground substance.
  - Identify the connective tissue cells.

3 Review the location and function for each connective tissue type in Tables 6.7–6.9.

4 Examine prepared microscope slides of the different connective tissue types. Use Figures 6.10–6.18 to help you locate the major structures in each tissue.

- collagen fiber
- connective tissue cells (fibroblasts or lymphocytes)
- elastic fiber

1 _____

2 _____

3 _____

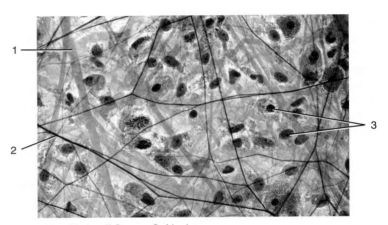

400×, Verhoeff Orange Safrin, human

FIGURE 6.10 Areolar connective tissue spread of mesentery.

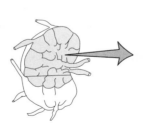

Lymph node

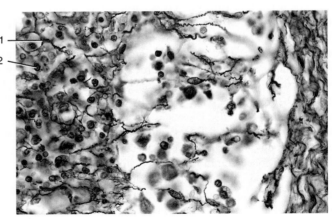

400×, human

- reticulocyte
- reticular fiber

1 _____

2 _____

FIGURE 6.11 Sectional view of reticular tissue of the lymph node.

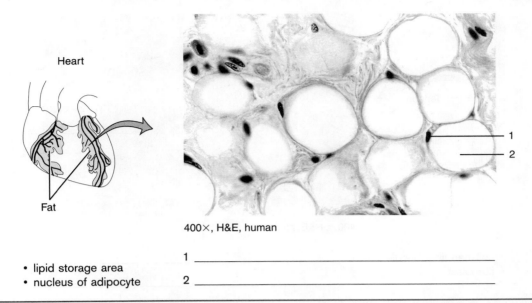

Heart

Fat

400×, H&E, human

- lipid storage area   1 _____
- nucleus of adipocyte   2 _____

FIGURE 6.12 Sectional view of adipose tissue.

TABLE 6.7 Location and Function of Loose Connective Tissues

TISSUE TYPE	LOCATION	FUNCTION
Areolar	Beneath all epithelial tissues	Binds epithelium to underlying tissues and allows nutrients to diffuse to epithelial cells
Reticular	Liver, spleen, lymph nodes	Forms support (framework) of these soft organs
Adipose	Under skin and surrounding organs	Stores lipids for fuel and thermal insulation; cushions organs

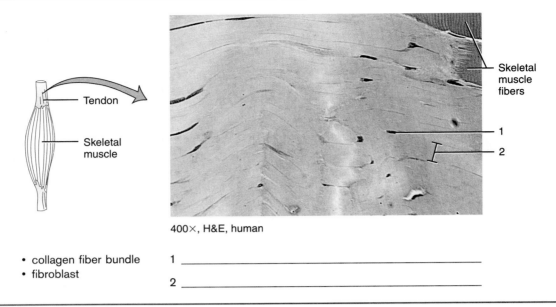

Tendon

Skeletal muscle

Skeletal muscle fibers

1

2

400×, H&E, human

- collagen fiber bundle
- fibroblast

1 _____

2 _____

FIGURE 6.13   **Sectional view of dense regular connective tissue forming tendons.**

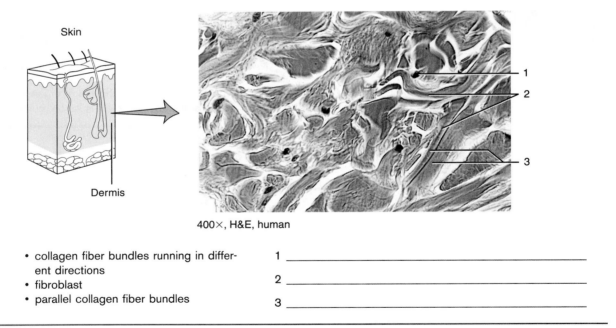

Skin

Dermis

1

2

3

400×, H&E, human

- collagen fiber bundles running in different directions
- fibroblast
- parallel collagen fiber bundles

1 _____

2 _____

3 _____

FIGURE 6.14   **Sectional view of dense irregular connective tissue in skin.**

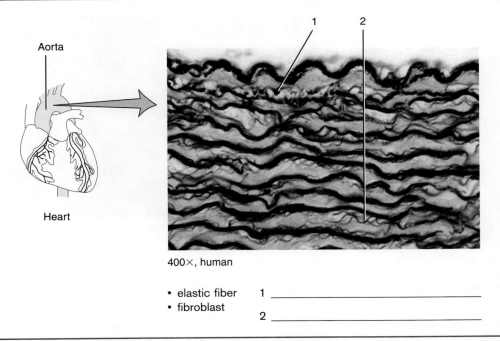

Aorta

Heart

400×, human

- elastic fiber   1 _____
- fibroblast   2 _____

FIGURE 6.15   Sectional view of elastic connective tissue in the wall of the aorta.

TABLE 6.8 Location and Function of Dense Connective Tissues

DENSE CONNECTIVE TISSUE TYPE	LOCATION	FUNCTION
Dense regular	Forms ligaments (connects bone to bone), tendons (connects muscle to bone), and aponeuroses (sheet-like tendons that connect muscle to muscle or muscle to bone)	Resists pulling forces at attachment points
Dense irregular	Skin	Resists pulling forces that would tear skin when skin is stretched
Elastic	Lungs, trachea, and bronchi	Allows respiratory organs to recoil after inhalation
	Aorta and other elastic arteries	Recoil of elastic tissue helps push blood through the cardiovascular system

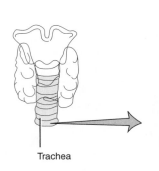

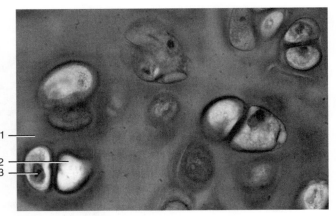

Trachea

400×, H&E, human

- extracellular matrix
- lacuna (la-KOO-na)
- nucleus of chondrocyte

1 _____

2 _____

3 _____

FIGURE 6.16 **Sectional view of hyaline cartilage in the wall of the trachea.**

TABLE 6.9 Location and Function of Cartilage

CARTILAGE	LOCATION	FUNCTION
Hyaline cartilage	Ends of long bones	Smooth surface that is resilient and reduces friction at joint
	Trachea and bronchi	Provides support and flexibility to ensure an open airway
	Anterior ends of ribs	Connects ribs to sternum with flexible joint (breastplate)
	Embryonic skeleton	Template for bone formation
Fibrocartilage	Intervertebral disks	Provides strength to disks that form joints between vertebrae and act as shock absorbers
	Cartilage pads in knee	Cushions surfaces of bones forming knee joints
	Pubic symphysis	Forms strong, flexible joint between hip bones
Elastic cartilage	External ear	Provides support and maintains shape of external ear
	Auditory tube	Provides support and elasticity to auditory tube as it changes diameter to equalize pressure in middle ear (ear pops)
	Epiglottis of larynx	Provides support and elasticity to epiglottis as it folds to block entrance to trachea while swallowing food and liquid

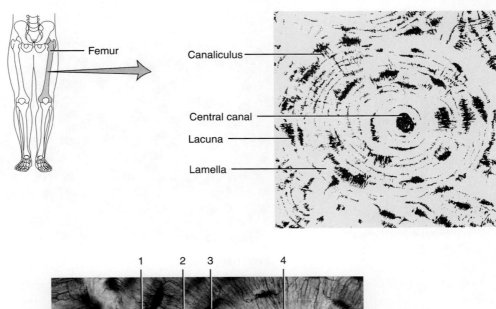

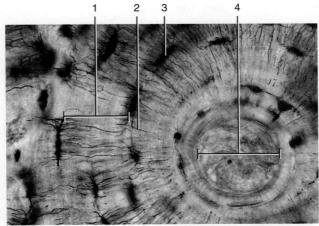

400×, unstained, human

- canaliculus
- central canal
- lamella
- lacuna

1 _____   3 _____

2 _____   4 _____

FIGURE 6.17   Sectional view of dried compact bone.

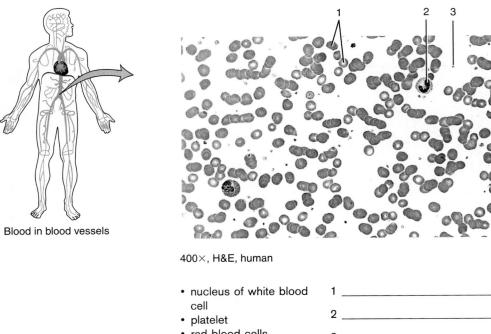

Blood in blood vessels

400×, H&E, human

- nucleus of white blood cell
- platelet
- red blood cells

1 _____

2 _____

3 _____

FIGURE 6.18 **Blood smear.**

C. MUSCLE TISSUE OVERVIEW

Muscle tissue is very cellular with most of the tissue consisting of muscle cells (Figure 6.19). All muscle tissues are highly vascularized and are innervated. Muscle cells are elongated cells called **fibers** that contain many microfilaments. These microfilaments cause the muscle cell to shorten when stimulated. This shortening or contraction causes movement. There are three types of muscle tissue—skeletal, cardiac, and smooth—with each type having a distinct appearance.

Skeletal and cardiac muscle cells exhibit **striations** (light and dark bands). Skeletal muscle cells are large, multinucleated, cylindrical cells. Cardiac muscle cells are smaller, branching cells with one centrally located nucleus per cell. Intercalated disks are dark bands where cardiac muscle cells connect end to end. Smooth muscle cells are small, spindle-shaped cells that are uninucleated and non-striated.

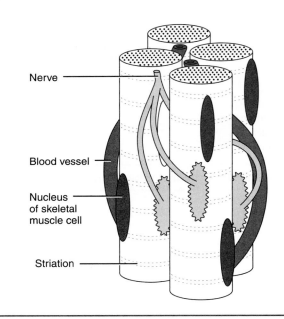

Nerve

Blood vessel

Nucleus of skeletal muscle cell

Striation

FIGURE 6.19 **Skeletal muscle tissue.**

ACTIVITY 3 MUSCLE TISSUE

1 Examine and color Figure 6.19, which depicts skeletal muscle tissue. Color a muscle fiber and its striations using contrasting colors.

2 Label and examine the photomicrographs of muscle tissue in Figures 6.20–6.22.

- Compare the size, shape, and number of nuclei of skeletal, cardiac, and smooth muscle fibers.
- Look for striations in skeletal and cardiac muscle fibers.

3 Review the location and function for each muscle tissue type in Table 6.10.

4 Examine prepared microscope slides of the different muscle tissue types.

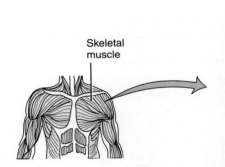

Skeletal muscle

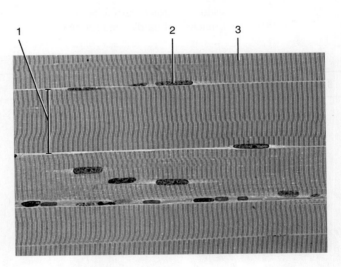

400×, H&E, human

- width of skeletal muscle fiber
- nucleus
- striation (stry-AY-tion)

1 _____

2 _____

3 _____

FIGURE 6.20 Sectional views of skeletal muscle fibers.

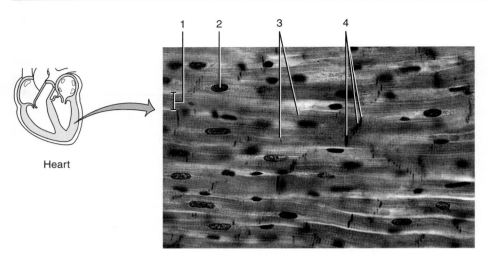

Heart

400×, iron H&E cardiac muscle tissue

- intercalated discs
- nucleus
- width of cardiac muscle fiber
- branches of cardiac muscle fiber

1 _____

2 _____

3 _____

4 _____

FIGURE 6.21 **Longitudinal view of cardiac muscle fibers.**

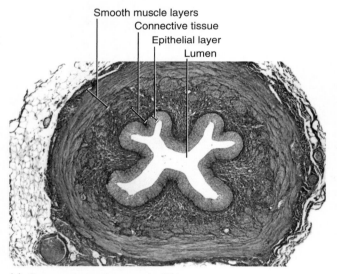

Smooth muscle layers
Connective tissue
Epithelial layer
Lumen

(a) Survey photomicrograph, 60×

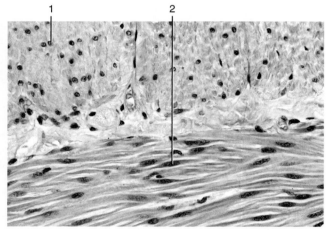

(b) Smooth muscle fibers in cross-section and longitudinal section, 400×

- nucleus of smooth muscle fiber in cross-section
- nucleus of smooth muscle fiber in longitudinal section

1 _____

2 _____

FIGURE 6.22 **Sectional view of smooth muscle in wall of the ureter.**

TABLE 6.10 Location and Function of Muscle Tissue Types

MUSCLE TISSUE	LOCATION	FUNCTION
Skeletal muscle	Attached to bones and skin	Movement of bones and skin
Cardiac muscle	Wall of heart	Movement of blood through cardiovascular system
Smooth muscle	Wall of digestive tract organs	Movement of food through digestive tract
	Walls of arteries and veins	Control of blood flow and blood pressure
	Walls of ureters, urinary bladder, and urethra	Movement of urine through urinary tract
	Intrinsic muscles of eye	Control pupil size

D. NERVOUS TISSUE OVERVIEW

Nervous tissue, which forms the brain, spinal cord, and nerves, is very cellular (Figure 6.23). Two basic categories of nervous tissue cells are **neurons** and **neuroglial** (*neuro* = nerve; *glia* = glue) **cells.** Neurons receive and send information, whereas neuroglia support the neurons and help them to function. Neurons have one or more processes (cellular extensions) that receive or send information as nerve impulses. **Dendrites** (*dendro* = tree) are processes that receive signals from sensory receptors or other neurons. An **axon** is another type of process that sends signals to other neurons, muscles, or glands. Neurons have only one axon, but may have many dendrites. The main part of the cell where the nucleus is located is called the **cell body.** In the following activity you will observe only **multipolar neurons,** which are neurons with many processes.

ACTIVITY 4 NERVOUS TISSUE

1 Examine and color Figure 6.23, which depicts nervous tissue. Color the multipolar neuron cell bodies, nuclei, and processes using contrasting colors.

2 Label and examine the photomicrograph of nervous tissue in Figure 6.24.
   • Compare the sizes of the neuron and neuroglia.
   • Observe all the sections of processes surrounding the neuron and neuroglial cells.

3 Examine a prepared microscope slide of a multipolar neuron.

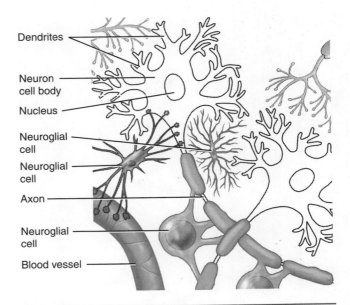

FIGURE 6.23 Nervous tissue.

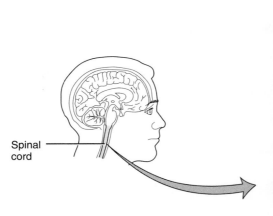

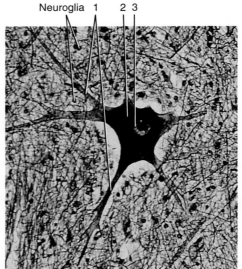

Spinal cord

LM 260×

- cell body of multipolar neuron
- nucleus
- processes

1 _____

2 _____

3 _____

FIGURE 6.24 **Photomicrograph of nervous tissue.**

REVIEWING YOUR KNOWLEDGE

A. Primary Tissue Structure and Function

Match the tissue characteristic to the primary tissue type. A characteristic could describe more than one primary tissue.

a. epithelial tissue b. connective tissue c. muscle tissue d. nervous tissue

____ 1. Cells contain processes that receive and gener- ate electrical signals to communicate with other cells.

____ 2. Cells of this tissue form glands.

____ 3. Tissue has elongated cells that shorten and cause movement.

____ 4. Tissue contains more extracellular matrix than cells.

____ 5. Cells form a barrier that controls passage of molecules.

____ 6. Some cell types contain striations.

____ 7. Primary tissue type(s) that exhibit(s) cellularity.

____ 8. Properties of cells determine function of tissue.

____ 9. Properties of extracellular matrix determine function of tissue.

B. Identification of the Four Primary Tissue Types

Identify primary tissue type in Figure 6.25.

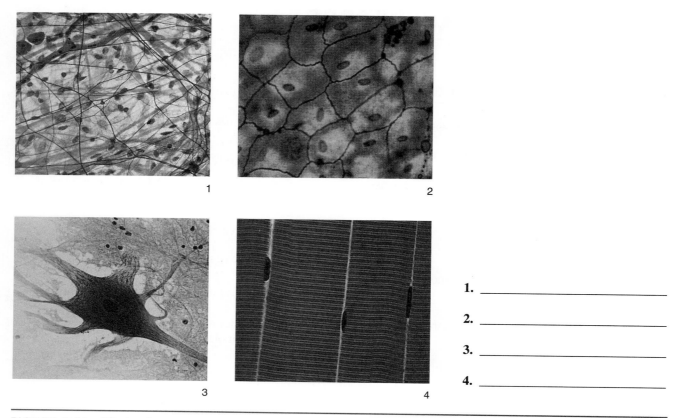

1. _____

2. _____

3. _____

4. _____

FIGURE 6.25   **Photomicrographs of primary tissues.**

C. Epithelial Tissues

Match each function to an epithelial tissue type. An epithelial tissue type could have more than one function.

a. absorption
b. diffusion
c. distensibility

d. filtration
e. movement of substances over the epithelial surface
f. protects underlying tissues in areas subjected to abrasion
g. secretion

_____ **1.** Stratified squamous epithelium lining the mouth.

_____ **2.** Simple squamous epithelium forming air sacs of the lung.

_____ **3.** Simple cuboidal epithelium forming kidney tubules.

_____ **4.** Ciliated pseudostratified epithelium lining the nasal cavities.

_____ **5.** Simple squamous epithelium forming mesothelium of serous membranes.

_____ **6.** Nonciliated simple columnar epithelium lining the stomach.

_____ **7.** Transitional epithelium lining the ureters.

D. Connective Tissues

Match each statement to its description. A connective tissue type could be used more than once.

a. areolar connective tissue
b. reticular connective tissue
c. adipose connective tissue
d. dense regular connective tissue
e. dense irregular connective tissue
f. elastic connective tissue

g. hyaline cartilage
h. bone
i. blood
j. collagen fibers
k. elastic fibers
l. reticular fibers

_____ 1. Connective tissue that contains elastic fibers and is found in the lung. This tissue allows the lungs to inflate during inhalation and return to their original shape after exhaling.

_____ 2. Connective tissue packed with parallel bundles of collagen fibers and found in tendons. This tissue resists pulling forces applied by muscles.

_____ 3. Connective tissue with a gelatinous matrix containing collagen fibers. This tissue is found in the tracheal wall; it provides support and prevents the trachea from collapsing.

_____ 4. Connective tissue found under covering and lining epithelium. Its extracellular matrix contains a loose arrangement of fibers, and its viscous ground substance facilitates the flow of interstitial fluid containing nutrients to epithelial tissues. It also cushions and supports epithelial tissues.

_____ 5. Connective tissue that forms a framework in the spleen, bone marrow, and lymph nodes. It contains finely branching collagen fibers.

_____ 6. Connective tissue that is packed with bundles of collagen fibers running in different directions. It is found in skin and allows skin to resist pulling forces from many different directions.

_____ 7. Fluid extracellular matrix used to transport substances throughout the body.

_____ 8. Connective tissue that contains a large number of lipid-storing cells. It is found throughout the body cushioning and insulating organs, and storing lipids for future energy needs.

_____ 9. Connective tissue with hardest extracellular matrix. Involved in protection and support.

_____ 10. Fibroblasts produce extracellular matrix for these connective tissue types.

_____ 11. Osteoblasts produce extracellular matrix.

_____ 12. Chondroblasts produce extracellular matrix.

_____ 13. Fibers not present unless injury occurs. Extracellular matrix not produced by cells present in this tissue.

_____ 14. Fine branching fibers with high tensile strength (resists pulling forces).

_____ 15. Thicker fibers with high tensile strength.

_____ 16. Resilient (can return to original shape after stretching).

E. Epithelial and Connective Tissue Identification

Identify the epithelial and connective tissue types in Figure 6.26.

1. _____

2. _____

3. _____

4. _____

5. _____

6. _____

7. _____

8. _____

9. _____

10. _____

11. _____

12. _____

13. _____

14. _____

15. _____

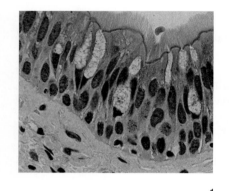

1

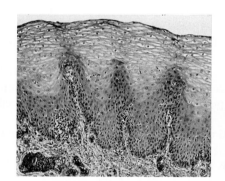

2

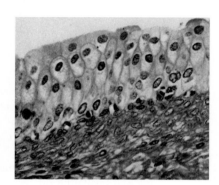

3

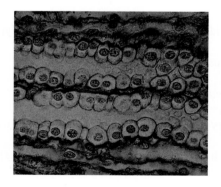

4

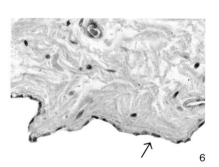

5

6

FIGURE 6.26 Identification of epithelial and connective tissue types.

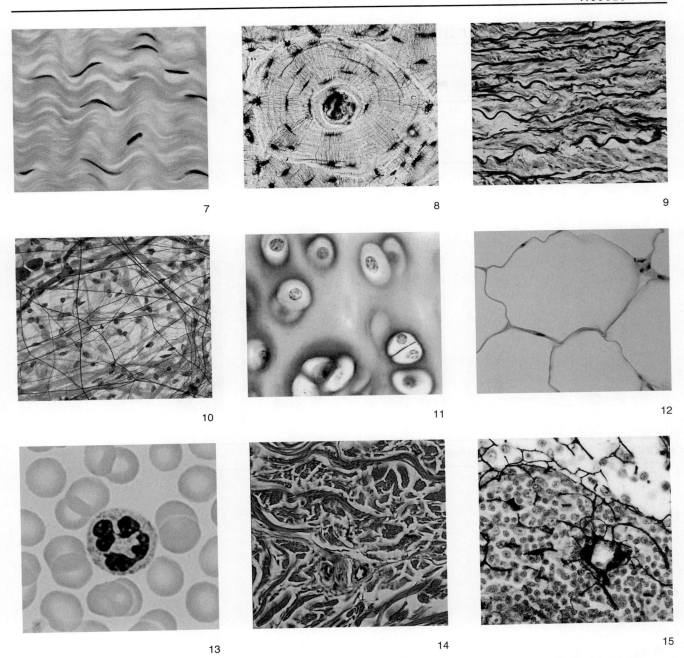

7

8

9

10

11

12

13

14

15

FIGURE 6.26 (Continued.)

F. Muscle Tissue Identification

Identify the muscle tissue types in Figure 6.27.

1. _____

2. _____

3. _____

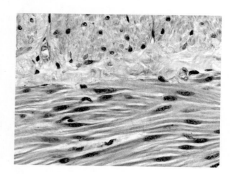 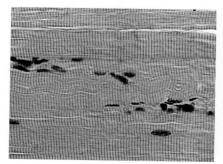 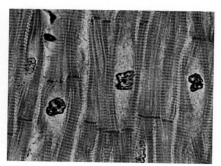

1   2   3

FIGURE 6.27   **Muscle tissue identification.**

THE INTEGUMENTARY SYSTEM

OBJECTIVES

After completing this exercise, you should be able to:

- Describe the structure and function of the integumentary system
- Identify the epidermis, dermis, and hypodermis on models, diagrams, or slides
- Describe the structure and function of the epidermal accessory structures
- Identify the accessory structures of the skin on models, diagrams, or slides

MATERIALS

- integumentary system model or chart
- prepared microscope slides: thick skin, thin skin with hair
- compound microscope and lens paper
- clean glass slides and cover slips
- ink pad for fingerprinting (or demonstration of fingerprints)
- 1 cm × 1 cm squares of bond paper (4 per student)
- Betadine-soaked cotton swabs
- surgical tape

The **integumentary system** (*inte-* = whole; *-gument* = body covering) consists of the skin, hair, nails, sweat and sebaceous glands, and associated smooth muscle and nervous tissue. This system provides a protective barrier for the body, contains sensory receptors, aids in the production of vitamin D, and is important in body temperature regulation.

A. MAJOR DIVISIONS OF THE SKIN

The skin has two major divisions: the epidermis and the dermis. The **epidermis** (*epi* = above), the outer layer of the skin, is composed of epithelial tissue. The **dermis** is the connective tissue layer that is firmly attached to and supports the epidermis, and connects it to the underlying hypodermis. Although it is not a part of the integumentary system, the **hypodermis** (*hypo* = below) or **subcutaneous layer** is located below the skin and is usually discussed with the skin.

1. Epidermis

The epidermis is a very thick epithelial layer in comparison with other epithelial layers of the body. The major cells of the epidermis are keratinocytes and melanocytes. **Keratinocytes** (*keratino* = horn-like; *cytes* = cells) comprise 90% of the cells of the epidermis. **Melanocytes** (*melano* = black) make up another 8%. The remaining 2% are immune cells and touch receptor cells.

The layers of the epidermis in order from deepest to most superficial are: stratum basale, stratum spinosum, stratum granulosum, stratum lucidum, and stratum corneum. The **stratum basale** (*strata* = layers; *basa-* = base) layer, a single row of cells, contains melanocytes that produce the pigment melanin and stem cells that divide to form new keratinocytes. As new keratinocytes are formed, they push the older cells toward the surface and undergo a process called keratinization. Keratinization produces tough, water repellant cells in the superficial layer.

The superficial layer, the **stratum corneum** (*corn-* = hard or hoof-like) is a very thick layer containing 25 to 30 rows of dead, squamous-shaped keratinocytes. This layer is tough and water-repellent. These cells continually slough off and are replaced by cells from the adjacent layer.

The epidermis differs in thickness in thin and thick skin. Thick skin is found on the palms of the hands and the soles of the feet and has all five strata. Thin skin, which covers the rest of the body, is missing the stratum lucidum layer.

2. Dermis

The dermis is highly vascularized connective tissue that allows nutrients to diffuse from the dermis into the epidermis. Hair follicles, sweat glands, and sebaceous glands that are derived from epithelial tissue and nerves are found in the dermis. The dermis also contains collagen fibers that provide the skin with strength and elastic fibers that provide elasticity and nerves.

Dermal papillae are finger-like projections of the dermis that extend into the epidermis. In the palms, fingers, soles, and toes, the dermal papillae cause genetically determined lines and whorls in the epidermis called **epidermal ridges** that increase surface area, friction, and grip. Sweat glands deposit their secretions onto these ridges, resulting in fingerprints when these ridges touch smooth surfaces.

ACTIVITY 1 MAJOR DIVISIONS OF THE SKIN

1 Label the skin structures in Figure 7.1.

2 Label the stratum basale and stratum corneum of the epidermis in Figure 7.2.

3 Identify the skin structures in Figures 7.1 and 7.2 on a model or chart.

4 Pronounce the terms as you point to them.

5 Examine a microscope slide of thick skin.

• Using the scanning or low-power objective, identify the epidermis, dermis, and hypodermis in Figure 7.1.

• Using the high-power objective, identify the stratum basale, stratum corneum, and dermal papillae of the epidermis in Figure 7.2. Observe the different cell shapes in the various layers of the epidermis.

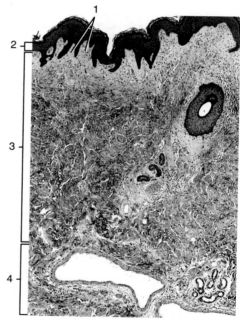

30×

• epidermis
• dermal papillae (puh-PILL-ee)
• dermis
• hypodermis

1 _____

2 _____

3 _____

4 _____

FIGURE 7.1 Photomicrograph of the skin.

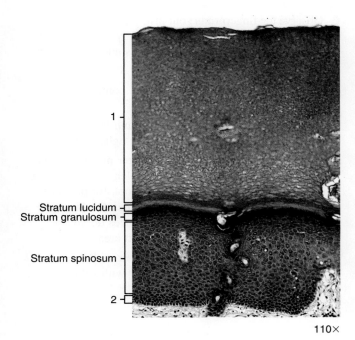

Stratum lucidum
Stratum granulosum

Stratum spinosum

110×

- stratum basale (bay-SAL)
- stratum corneum (kor-NEE-um)

1 _____

2 _____

FIGURE 7.2 Photomicrograph of the epidermal layers in thick skin.

B. ACCESSORY STRUCTURES OF THE SKIN

The accessory structures of the skin include hair, hair follicles, nails, sweat glands, and oil glands. All of these are derived from epithelial tissues but are located in connective tissue.

1. Hair and Hair Follicles

Hairs, which help protect and insulate, are found over most of the body and consist of dead, keratinized epithelial cells. They have two sections: the **shaft,** which is located in the epidermis and projects from the skin surface, and the **root,** which is embedded in the dermal layers of the skin. The **hair follicle,** located in the dermis and surrounding the hair root, is formed from epidermal layers that project into the dermis. The expanded base of the hair follicle, the **hair bulb,** contains stem cells that produce new hair cells. Attached to the connective tissue around the hair follicle is a bundle of smooth muscle cells called the **arrector pili** (*arrect* = to raise). Contraction of these muscles move the hair from its normal angle to a 90° angle (perpendicular) with skin surface, elevating the skin surrounding the hair shaft and causing "goosebumps."

Examination of the hair in cross- or longitudinal section shows three layers. The **cuticle** (thin, outer layer) is composed of dead, flattened, keratinized cells that overlap each other like shingles on a roof. Split ends occur when the free ends of these cells are pulled away from each other. Cells of the **cortex** (middle layer) and the **medulla** (center layer) contain pigment granules that give hair its color.

2. Sudoriferous and Sebaceous Glands

Sudoriferous glands (*sudori* = sweat; *-ferous* = bearing) or sweat glands are located in the dermis and secrete a watery substance for body temperature regulation and excretion. There are two types of sudoriferous glands: eccrine glands and apocrine glands. **Eccrine glands** (*eccrine* = sweating outwardly) are the most common type of sweat gland and are found on most areas of the body. Ducts from these glands deposit secretions called sweat on the skin. **Apocrine glands** are found only in the axilla, genital area, and pigmented area around the nipples (areolae). Apocrine glands produce a secretion similar to sweat, but it is more viscous. This secretion, which is deposited on the distal end of the hair root, is odorless until broken down by bacteria. **Ceruminous glands** (*ceri-* = wax) are modified sweat glands found in the ear canal. **Mammary glands** are modified sweat glands found in the breasts.

Sebaceous glands (*sebace-* = greasy) or oil glands are found surrounding hair follicles and deposit an oily substance called sebum, that lubricates the skin and hair, onto the hair shaft.

ACTIVITY 2 **HAIR, HAIR FOLLICLES, SEBACEOUS GLANDS, AND SUDORIFEROUS GLANDS**

1 Label the diagram of the skin and accessory structures in Figure 7.3.

2 Point to the structures on a model or chart of the integumentary system.

3 Pronounce the names of the structures as you point to them.

4 Label the photomicrograph of the skin in Figure 7.4.

5 Examine a slide of hairy skin, and identify a hair bulb, hair follicle, hair root, and sebaceous gland.

6 Observe the fingerprints your instructor has as a demonstration or use an ink pad to make your fingerprints. Examine the fingerprints of several students to observe the variety of whorl patterns.

7 Examine a strand of hair with the compound microscope.

- Pull out one strand of scalp hair and lay the hair horizontally on the stage over the light source.

- Observe the dark cuticle and lighter medulla of the hair shaft with the 4 × objective lens. Switch to the low- and high-power lenses and look for the cells of the cuticle overlapping each other. Observe the root end of the hair. What differences do you see between the root end and the shaft area?

- If someone is willing to donate hair with "split ends," compare the cuticle of hair with split ends with undamaged hair.

- apocrine sweat gland
- arrector pili muscle
- eccrine sweat gland
- hair bulb
- hair follicle
- hair root
- hair shaft
- sebaceous gland

1 _____

2 _____

3 _____

4 _____

5 _____

6 _____

7 _____

8 _____

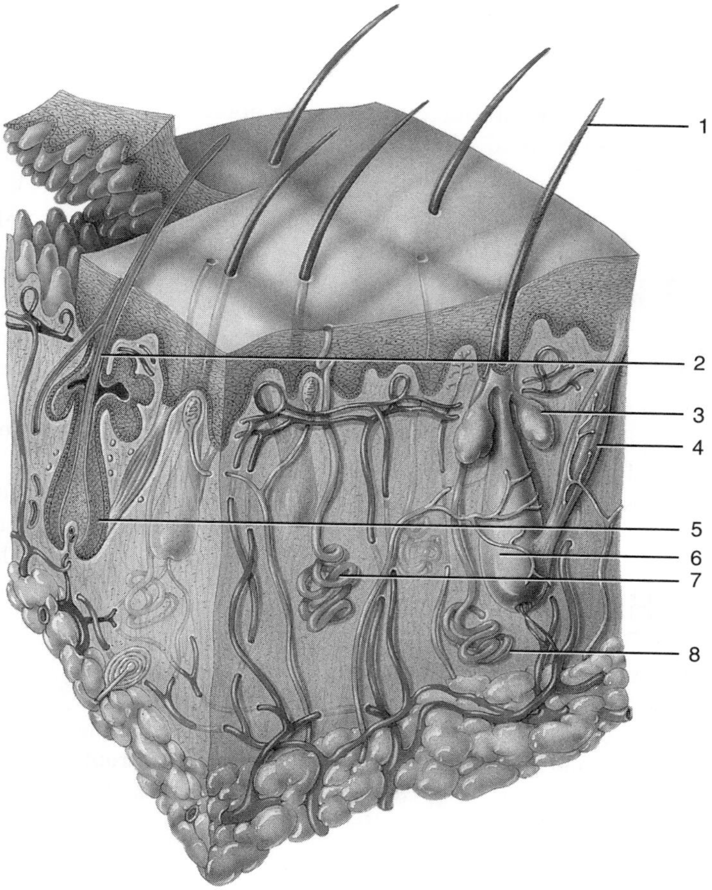

FIGURE 7.3 **Diagram of the skin and accessory structures.**

8 Compare concentration of sweat glands in different areas of the body.

- Obtain 4 squares of bond paper (each square 1 cm × 1 cm), a Betadine-soaked cotton swab, and tape.
- Find areas on your forehead, anterior forearm, palm, and anterior leg that do not have large crease lines.
- For each location, apply Betadine to an area slightly larger than the paper square.
- Securely tape a paper square over the Betadine-covered area at each location.

- After 20 minutes, remove the tape and paper squares, and count the blue-black dots on each square. Each blue-black dot is an active sweat gland.
- Record the number of dots per cm² in Table 7.1.
- Pool the class data, calculate the average eccrine gland density, and record the value in Table 7.1.

9 Answer the discussion questions with your lab partners.

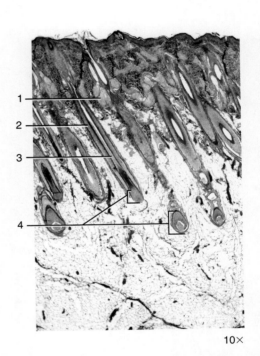

- hair bulbs
- hair follicle
- hair root
- sebaceous gland

1 _____

2 _____

3 _____

4 _____

10×

FIGURE 7.4   Photomicrograph of the skin and accessory structures.

TABLE 7.1 Eccrine Sweat Gland Density

BODY AREA	ECCRINE SWEAT GLAND DENSITY (NO. OF GLANDS/CM²)	
	INDIVIDUAL VALUE	CLASS AVERAGE
Forehead		
Anterior forearm		
Palm		
Anterior leg		

❓DISCUSSION QUESTIONS: ECCRINE GLAND DENSITY

1 Which body area had the highest density of eccrine glands and which had the lowest?

2 Compare eccrine gland density with the amount of sweat that appears on these body areas during exercise or nervousness.

REVIEWING YOUR KNOWLEDGE

A. Identification of Skin Layers and Accessory Structures

Label the skin layers and accessory structures in Figure 7.5.

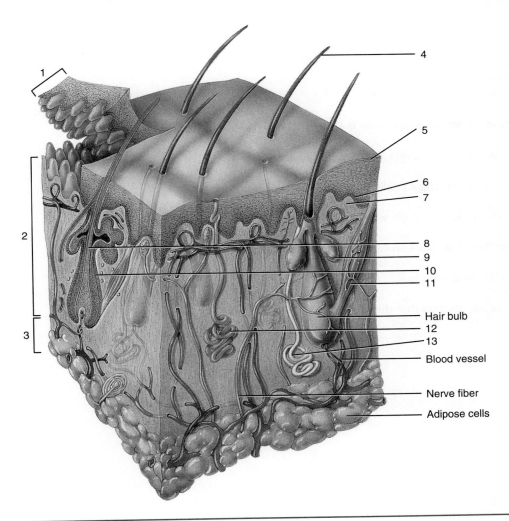

FIGURE 7.5   Skin layers and accessory structures.

1. _____   4. _____

2. _____   5. _____

3. _____   6. _____

7. _____   11. _____

8. _____   12. _____

9. _____   13. _____

10. _____

B. Skin Layers

Choose the skin layer or cell type that fits the description.

a. dermal papillae
b. dermis
c. epidermis
d. keratinocytes
e. melanocytes
f. stratum basale
g. stratum corneum

_____ 1. Layer of epidermis where there is the most rapid cell division

_____ 2. Vascular connective tissue layer beneath epidermis of skin

_____ 3. Tough, water-repellent epidermal layer; contains dead squamous-shaped cells

_____ 4. Cells that produce keratin which forms the waterproofing of skin

_____ 5. Projections of dermis that cause epidermal ridges

_____ 6. Epithelial layer of skin

_____ 7. Produce the pigment that shields cells from UV radiation

C. Accessory Structures of the Skin

Select the choice that best fits the description.

a. apocrine sweat gland
b. arrector pili muscle
c. eccrine sweat gland
d. hair bulb
e. hair follicle
f. hair root
g. hair shaft
h. sebaceous gland

_____ 1. Sudoriferous glands located in axillary and genital areas; become active after puberty

_____ 2. Epithelial structure that surrounds hair root

_____ 3. Secretes sebum onto hair and skin

_____ 4. Part of hair that contains stem cells that produce new hair cells

_____ 5. Sudoriferous glands that deposit sweat onto epidermal ridges causing fingerprints

_____ 6. Part of hair in epidermis and extending beyond skin surface

_____ 7. Moves hair shaft perpendicular to skin and causes goosebumps

_____ 8. Part of hair within dermis

D. Application

1. Explain how dandruff is formed.

2. Name the layer of skin that "peels" off after a minor sunburn.

3. Permanent tatoos are made by injecting pigment into the skin. Into which layer of the skin is the pigment injected, the epidermis or dermis? (You can answer this without research, just apply your knowledge).

BONE STRUCTURE

OBJECTIVES

After completing this exercise, you should be able to:

- Identify bones as either long, short, flat, irregular, or sesamoid
- Describe the gross structure of a long bone
- Describe the chemical composition of bone

MATERIALS

- human long bone cut longitudinally and transversely
- fresh animal bones sectioned longitudinally
- 1-inch cross-section cut from fresh long bone
- dissecting tray, blunt probe, forceps, and disposable gloves
- chicken leg bone baked at 250°F for 2 hours or more
- chicken leg bone soaked in an acidic solution (vinegar for 1 week, cola for 3–4 days, or nitric acid for 3–4 days)
- untreated chicken leg bone
- disarticulated bones, for bone classification

Bones are composed of a complex arrangement of several tissues. A typical bone has two types of osseous tissue: compact and spongy. In addition, bones contain dense connective tissue, cartilage, adipose tissue, tissue that forms blood cells, nerves, and blood vessels. The study of bone structure and disorders is called **osteology** (*oste-* = bone; *-ology* = study of).

A. CLASSIFICATION OF BONES

The bones of the human body have different shapes and distinct gross anatomical features. Bones are placed in five classifications according to their shapes: long, short, flat, irregular, and sesamoid. **Long bones** are longer than they are wide, **short bones** are almost equal in length and width, **flat bones** are thin and relatively flat but may be curved, **irregular bones** are self-explanatory and do not easily fit into any of these categories, and **sesamoid bones** (*sesamoid* = sesame seed) are small bones that develop in tendons for protection against wear and tear. The patella (kneecap) is the largest sesamoid bone, and some people have additional small sesamoid bones.

ACTIVITY 1 CLASSIFICATION OF BONES ACCORDING TO SHAPE

1 Identify the types of bones in Figure 8.1 according to shape.

2 Classify each of the bones your instructor has displayed according to shape.

B. GROSS FEATURES OF LONG BONES

Each enlarged proximal and distal end of a long bone is called an **epiphysis** (*epi-* = above, over; *physis* = growing). The epiphysis contains spongy bone covered by a thin layer of compact bone. The middle shaft area composed of compact bone is called the **diaphysis** (*dia-* = through). The area of the epiphysis where the bone grows in length is called the **epiphyseal plate** in a growing bone and the **epiphyseal line** in a bone that is no longer growing. The cavities within the spongy bone of the epiphyses (plural) contain **red marrow**, a tissue that produces blood cells. **Articular cartilage** (*articul-* = joint), composed of hyaline cartilage, covers both epiphyses, and the rest of the bone exterior is covered with a tough, connective tissue membrane, the **periosteum** (*peri-* = around; *osteo-* = bone). The hollow center of the bony diaphysis is called the **medullary cavity** (*medulla-* = marrow, pith), and a small amount of spongy bone is found in this cavity. The medullary cavity is lined with a connective tissue mem-

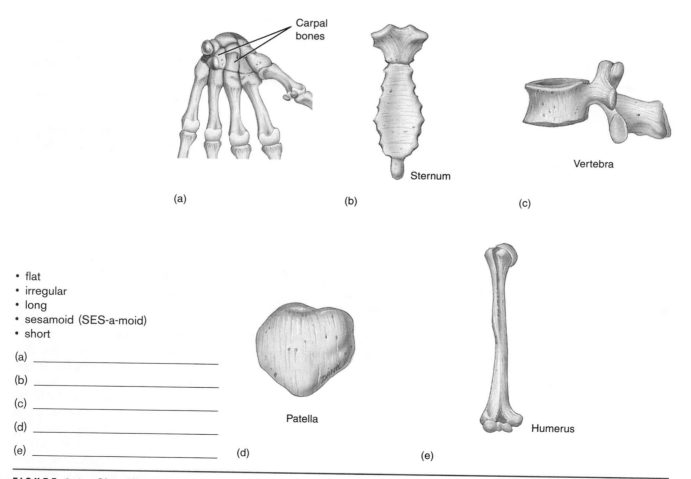

Carpal bones

(a)

Sternum

(b)

Vertebra

(c)

• flat
• irregular
• long
• sesamoid (SES-a-moid)
• short

(a) _____

(b) _____

(c) _____

(d) _____

(e) _____

Patella

(d)

Humerus

(e)

FIGURE 8.1 Classification of bones.

brane called the **endosteum** (*endo-* = within) and is filled with a fatty substance called **yellow marrow.** The endosteum also lines the cavities within the spongy bone of the epiphyses. Both the periosteum and the endosteum contain **osteoblasts** (*-blast* = builder) and **osteoclasts** (*-klasis* = breaking) for bone formation, as well as bone tissue repair.

3 Use a dissecting microscope to observe the medullary cavity with spongy bone at a higher magnification.

4 Pronounce the terms as you point to them on the bone.

ACTIVITY 2 GROSS FEATURES OF A LONG BONE

1 Label Figure 8.2.

2 Identify the items from Figure 8.2 on a long bone that is partially sectioned longitudinally and transversely.

> **SAFETY NOTE:** *Wear gloves when using fresh tissue! Wash hands thoroughly with soap and water at the conclusion of this experiment.*

- articular cartilage
- compact bone
- diaphysis (die-AF-ih-sis)
- distal epiphysis (e-PIF-ih-sis)
- endosteum (en-DOS-tee-um)
- epiphyseal line (ep-i-PHY-zee-al or ee-PIF-ih-seal)
- medullary cavity (MED-yoo-lar-y)
- periosteum (peri-OS-tee-um)
- proximal epiphysis
- spongy bone
- yellow marrow

1 _____
2 _____
3 _____
4 _____
5 _____
6 _____
7 _____
8 _____
9 _____
10 _____
11 _____

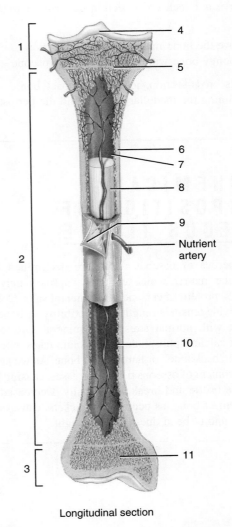

Longitudinal section

FIGURE 8.2 **Features of an adult long bone.**

ACTIVITY 3 FRESHLY CUT SECTION OF LONG BEEF BONE

1 Using gloves, examine a fresh cut section of an animal bone.

2 Identify as many features as you can from Activity 2. Is the articular cartilage dull or shiny?

Is the periosteum dull or shiny?

3 Note the consistency of the bone marrow.

4 Find the epiphyseal line.

5 Use a blunt probe and forceps to loosen the periosteum, ligaments, or tendons.

6 Observe a 1-inch cross-section cut from a fresh long bone.

7 Remove the bone marrow from this section to observe the spongy bone with a dissecting microscope.

8 In this cross-section, identify compact bone, spongy bone lining the medullary cavity, and the periosteum.

C. CHEMICAL COMPOSITION OF OSSEOUS TISSUE

The properties of osseous tissue are determined by its extracellular matrix which contains approximately 25% water, 25% protein fibers, and 50% mineral salts. Collagen fibers provide tensile strength and flexibility so bone does not break with normal stress. The mineral salts consist mainly of calcium phosphate and calcium carbonate salts, giving the "backbone" or hardness to bone. As we age, the collagen content of osseous tissue decreases, causing bones to become brittle and break more easily. Decreased mineral content of bone, as occurs with rickets, causes bones to be soft and to bend due to body weight.

ACTIVITY 4 CHEMICAL COMPOSITION OF BONE

1 With your group, examine a chicken leg bone baked at about 250°F for a minimum of 2 hours or until brittle. Compare what happens when you try to bend a treated and an untreated bone.

a. What has happened to the flexibility of this bone?

b. Is this bone more or less brittle than an untreated bone?

c. What substance from this bone has been affected?

d. How does this affect bone breakage?

2 Examine a chicken leg bone that has been soaked in an acidic solution. Try bending the bone and compare this with an untreated bone and the baked bone.

a. What has happened to the flexibility of this bone?

b. What has happened to the hardness of this bone?

c. What substance has been leached from this bone?

d. What clinical disorder with bone softness is this activity simulating?

REVIEWING YOUR KNOWLEDGE

A. Gross Features of Long Bones

1. What area of the long bone is covered with cartilage? _____

2. What type of cartilage is articular cartilage? _____

3. What area (epiphysis or diaphysis) is made up of a *thin* layer of compact bone and *thick* spongy bone?

_____

4. What area (epiphysis or diaphysis) is made up of a *thick* layer of compact bone and a *very thin* layer of spongy

bone? _____

5. The long bone in Figure 8.2 is from an adult. If this figure was from a child, what structure would be present instead

of the epiphyseal line? _____

B. Chemical Composition of Bone

Complete the sentences with the correct word or words.

_____ **1.** The hardness of bone is due to _____.

_____ **2.** The flexibility and tensile strength of bone are due to _____.

_____ **3.** What type of a macromolecule (carbohydrate, lipid, protein) is collagen?

_____ **4.** A bone that has the collagen removed is flexible or inflexible?

_____ **5.** A bone that has calcium removed is flexible or inflexible?

C. Application

1. Observe the bones in Figure 8.3(a) and 8.3(b). Identify the child's hand and the adult's hand.

(a) _____ (b) _____

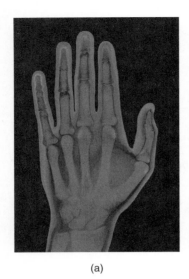 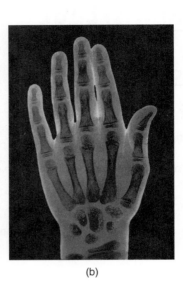

(a)   (b)

FIGURE 8.3 **X-rays of the child and adult hand.**

2. Observe the scanning electron micrographs (SEMs) of bone tissue in Figure 8.4(a) and 8.4(b). Identify the normal bone and the osteoporotic bone.

(a) _____ (b) _____

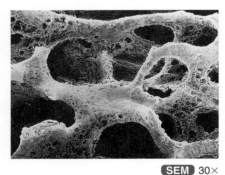 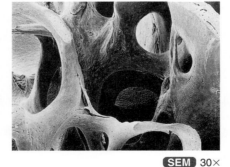

SEM 30×   SEM 30×

(a)   (b)

FIGURE 8.4 **Normal and osteoporotic bone tissue.**

3. Is the bone tissue in Figure 8.4 spongy bone or compact bone? _____

AXIAL SKELETON

OBJECTIVES

After completing this exercise, you should be able to:

- Identify the three main parts of the axial skeleton
- Identify major bones and selected bone surface markings and sutures of the axial skeleton
- Identify the location of the paranasal sinuses on a model or chart
- Identify major fontanels in the fetal skull
- Identify the parts of a typical vertebra and identify the three individual types of vertebrae
- Compare and contrast the normal and abnormal curvatures of the vertebral column

MATERIALS

- articulated skeleton
- disarticulated skeleton
- individual skulls, Beauchene skull, and fetal skull
- small straws or pipe cleaners (for pointers)
- colored pencils

There are 206 bones in the adult skeleton, which can be separated into the **axial** and the **appendicular** divisions. The **axial skeleton** is composed of 80 bones located along a vertical line, the longitudinal axis (line) of the body. Its bones support and protect the organs of the head, neck, and torso. The **appendicular skeleton** (*appendere* = to hang upon) is composed of 126 bones that are designed for a variety of movements and form the upper and lower limbs. The appendicular skeleton will be studied in Exercise 10.

The three main parts of the axial skeleton are the **skull, vertebral column,** and **thoracic cage.** Smaller bones included in this division are the **ear ossicles** (bones).

A. THE SKULL AND SUTURES

The **cranium** (cranial bones) surrounds and protects the brain and houses organs of hearing and equilibrium. **Facial bones** provide the shape of the face, house the teeth, and provide attachments for the muscles of facial expression. Other major features of the skull include **sutures, bone markings, paranasal sinuses, nasal septum, hard palate,** and **fontanels** (fetal skull). The **hyoid bone** is studied here because of its location, but it is not a skull bone.

1. Cranial Bones

There are a total of eight large bones in the cranium, two paired and four single bones.

- **(2) parietal bones** (*paries* = wall)—superior lateral wall
- **(2) temporal bones**—inferior lateral wall; house organs of inner ear
- **(1) frontal bone**—anterior portion
- **(1) occipital bone** (*occipit-* = atlas)—posterior wall
- **(1) sphenoid bone** (*sphen-* = wedge; *-eidos* = form)—floor posterior to ethmoid
- **(1) ethmoid bone** (*ethmos* = sieve)—floor and anterior wall

2. Sutures

Sutures are immovable joints between cranial and facial bones. The four main sutures in the skull are the:

- **coronal** (*corona* = crown)—joins frontal and parietal bones
- **sagittal** (*sagitta-* = arrow)—joins parietal bones
- **lambdoid** (shape of the Greek letter *lambda*)—joins both parietal bones with occipital bone
- **squamous**—joins temporal and parietal bones

3. Facial Bones

There are a total of 14 **facial bones,** six paired and two single bones.

- **(2) maxillae** (*mala* = jaw)—upper jaw and hard palate
- **(2) zygomatic bones** (*zygoma* = yoke or bar)—cheek bone
- **(2) lacrimal bones** (*lacri-* = tear)—orbit of eye near nasal bone
- **(2) nasal bones**—bridge of nose
- **(2) inferior nasal conchae** (*concha* = shell or scroll-shaped) or turbinate—lateral wall of nasal cavity
- **(2) palatine bones** (*palatum* = palate)—posterior part of hard palate
- **(1) mandible** (*mandere* = to chew)—lower jaw
- **(1) vomer** (*vomer* = plowshare)—floor of nasal cavity and inferior portion of nasal septum

4. Selected Bone Surface Markings of the Skull

Bones of the skull are not totally smooth on the exterior but have bumps, depressions, grooves, holes, and projections. Bone markings have specific names and distinct purposes, and are classified according to their function. Several bone markings found on the skull are listed in Table 9.1 along with their descriptions and functions. Selected bones and their markings are listed below.

Cranial Bones: Selected Bone Surface Markings

Temporal bones—Each temporal bone has one of each structure listed below:

- **external auditory meatus** (*meatus* = passageway)—tube-like openings for the ear canal
- **mandibular fossa**—depression for articulation with condylar process (mandibular condyle) to form the *temporomandibular joint (TMJ)*
- **mastoid process** (*mastoid* = breast-like)—rounded projection posterior to external auditory meatus; attachment for muscles

TABLE 9.1 Selected Bone Markings of the Axial Skeleton

MARKING	DESCRIPTION	PURPOSE
1. Foramen (for-AY-men) (plural is foramina)	Opening or hole	Opening for blood vessels and nerves
2. Fossa (FOS-sa)	Shallow depression	Muscle attachment or articulation (joint)
3. Meatus (me-AY-tus)	Tube-like opening	Tubular passageway for blood vessels and nerves
4. Condyle (CON-dile)	Smooth, rounded articular process	Articulation

- **styloid process** (*stylo-* = point)—long, thin projection on inferior skull surface; attachment for muscles and ligaments of tongue and neck
- **zygomatic arch**—formed from projection of the temporal bone and zygomatic bone
- **carotid foramen (canal)**—foramen for carotid artery

Occipital bone

- **foramen magnum** (*magnum* = large)—opening connecting lower brain to spinal cord
- **occipital condyles**—two rounded processes that articulate with the atlas (C1)

Ethmoid bone

- **cribriform plates** (*cribr-* = sieve)—one on either side of crista galli; form roof of nasal cavity
- **crista galli** (*crist-* = crest; *galli* = rooster)— projection for attachment of membranes covering brain
- **perpendicular plate**—forms the upper part of nasal septum
- **middle nasal conchae**—scroll-like projections on each lateral wall of nasal cavity
- **superior nasal conchae**—scroll-like projections on each lateral wall of nasal cavity

Sphenoid bone

- **optic foramina**—openings for optic nerve (II)
- **sella turcica** (*sella* = saddle; *turcica* = Turkish)— bony projection that surrounds pituitary gland

Facial Bones: Selected Bone Surface Markings

Maxillae

- **alveoli** (*alve-* = socket)—teeth sockets for upper teeth

Mandible

- **alveoli**—teeth sockets for lower teeth
- **condylar processes (mandibular condyles)**—two superior, rounded processes that articulate with temporal bone at the mandibular fossa; together they form the TMJ

> **NOTE:** *BE ACCOUNTABLE FOR THE CARE OF SKULLS AND BONES. Please be careful not to use pencils, pens, or markers as pointers while you are studying the skull and other bones. Thin bones can be broken, and other bones can be permanently marked. Your instructor will provide a small straw, pipe cleaner, or other suitable pointer.*

ACTIVITY 1 CRANIAL AND FACIAL BONES, SUTURES, AND BONE MARKINGS

1 Label the cranial bones and facial bones, sutures, and selected bone markings on the photographs in Figures 9.1–9.4.

2 Using contrasting colors, color the bones of the skull in the drawings of Figures 9.1–9.4.

3 Identify these structures on an articulated skull (and a Beauchene skull, if available).

4 Pronounce the terms as you point to them on a skull.

5 Using two straws, poke one into each orbit of the eye on the skull and through an optic foramen. Observe the straws in the superior view of the cranial floor and notice that the straws cross above the sella turcica. This simulates the crossing of the optic nerves called the optic chiasma.

6 Palpate the cranial and facial bones on your own skull, your zygomatic arch, and your mastoid process.

7 Palpate your TMJ by putting your fingers on either side of your jawbone just anterior to the ears. Open and close your mouth to feel this joint.

> **CLINICAL NOTE:** *A dislocated or defective TMJ has a history of causing clinical problems characterized by pain, tenderness, dysfunction, and clicking noises.*

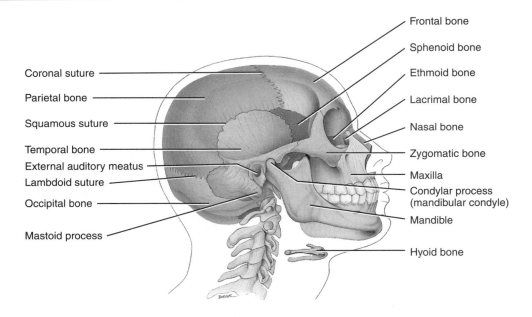

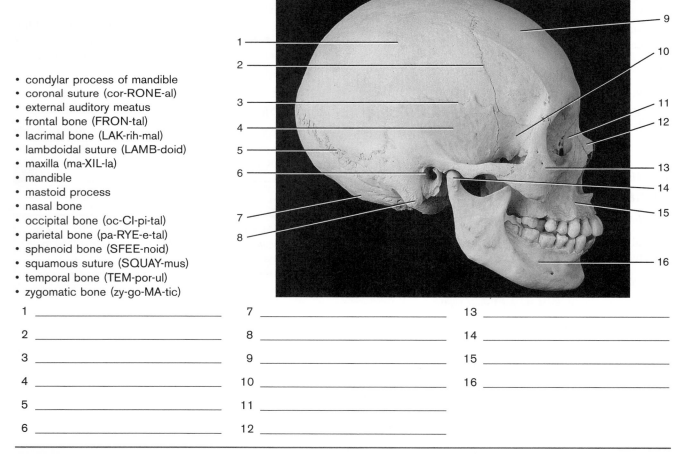

- condylar process of mandible
- coronal suture (cor-RONE-al)
- external auditory meatus
- frontal bone (FRON-tal)
- lacrimal bone (LAK-rih-mal)
- lambdoidal suture (LAMB-doid)
- maxilla (ma-XIL-la)
- mandible
- mastoid process
- nasal bone
- occipital bone (oc-CI-pi-tal)
- parietal bone (pa-RYE-e-tal)
- sphenoid bone (SFEE-noid)
- squamous suture (SQUAY-mus)
- temporal bone (TEM-por-ul)
- zygomatic bone (zy-go-MA-tic)

1 _____	7 _____	13 _____
2 _____	8 _____	14 _____
3 _____	9 _____	15 _____
4 _____	10 _____	16 _____
5 _____	11 _____	
6 _____	12 _____	

F I G U R E 9 . 1 Lateral view of skull.

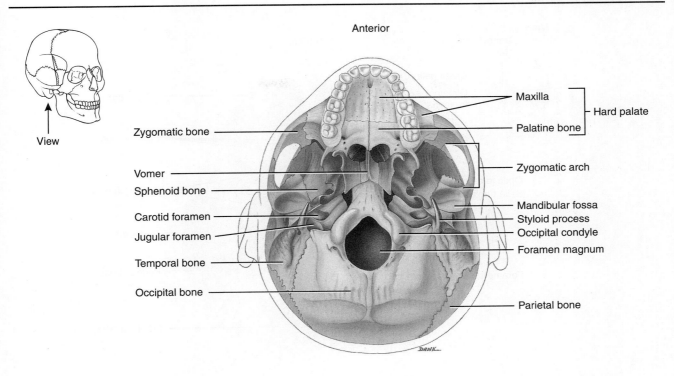

Anterior

View

Maxilla

Hard palate

Zygomatic bone

Palatine bone

Zygomatic arch

Vomer

Sphenoid bone

Mandibular fossa

Carotid foramen

Styloid process

Jugular foramen

Occipital condyle

Foramen magnum

Temporal bone

Occipital bone

Parietal bone

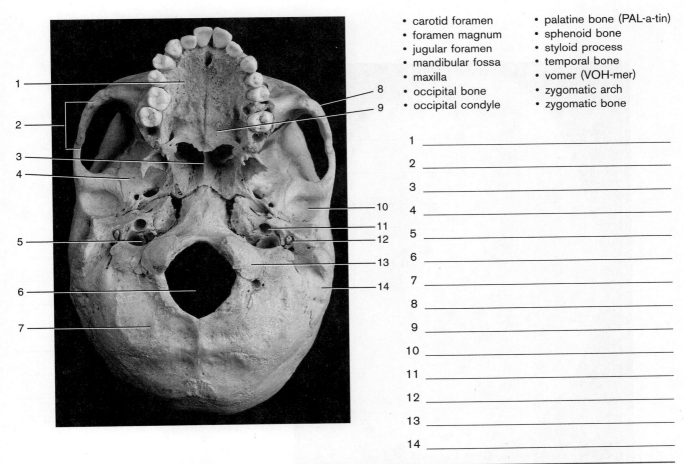

- carotid foramen
- foramen magnum
- jugular foramen
- mandibular fossa
- maxilla
- occipital bone
- occipital condyle

- palatine bone (PAL-a-tin)
- sphenoid bone
- styloid process
- temporal bone
- vomer (VOH-mer)
- zygomatic arch
- zygomatic bone

1 _____
2 _____
3 _____
4 _____
5 _____
6 _____
7 _____
8 _____
9 _____
10 _____
11 _____
12 _____
13 _____
14 _____

FIGURE 9.2   Inferior view of skull.

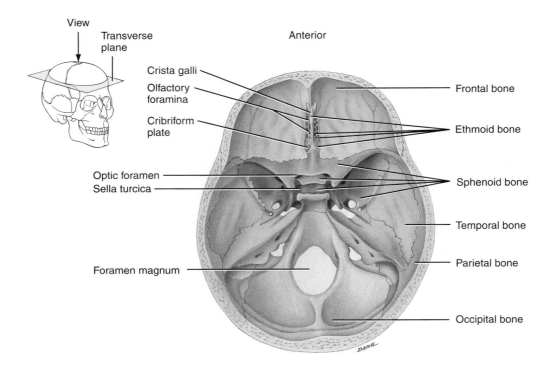

View
Transverse plane
Anterior

Crista galli — Frontal bone
Olfactory foramina
Cribriform plate — Ethmoid bone

Optic foramen — Sphenoid bone
Sella turcica

— Temporal bone

Foramen magnum — Parietal bone

— Occipital bone

Posterior

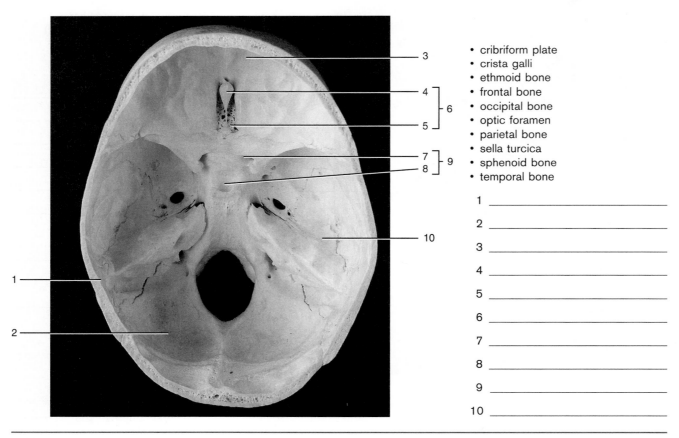

- cribriform plate
- crista galli
- ethmoid bone
- frontal bone
- occipital bone
- optic foramen
- parietal bone
- sella turcica
- sphenoid bone
- temporal bone

1 _____
2 _____
3 _____
4 _____
5 _____
6 _____
7 _____
8 _____
9 _____
10 _____

FIGURE 9.3 **Superior view of floor of cranium.**

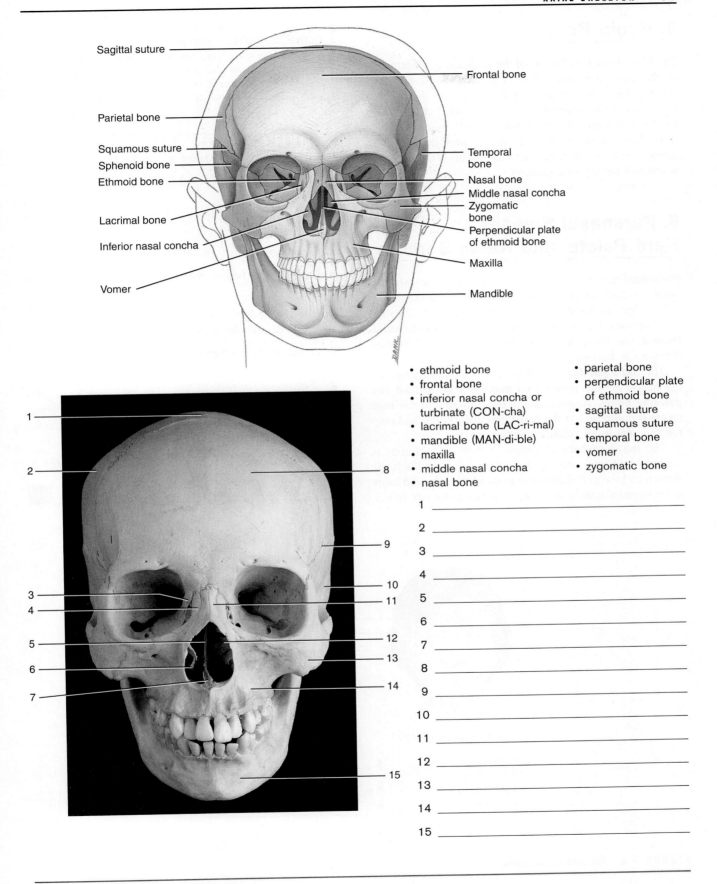

Sagittal suture

Parietal bone

Squamous suture

Sphenoid bone

Ethmoid bone

Lacrimal bone

Inferior nasal concha

Vomer

Frontal bone

Temporal bone

Nasal bone

Middle nasal concha

Zygomatic bone

Perpendicular plate of ethmoid bone

Maxilla

Mandible

- ethmoid bone
- frontal bone
- inferior nasal concha or turbinate (CON-cha)
- lacrimal bone (LAC-ri-mal)
- mandible (MAN-di-ble)
- maxilla
- middle nasal concha
- nasal bone

- parietal bone
- perpendicular plate of ethmoid bone
- sagittal suture
- squamous suture
- temporal bone
- vomer
- zygomatic bone

1 _____

2 _____

3 _____

4 _____

5 _____

6 _____

7 _____

8 _____

9 _____

10 _____

11 _____

12 _____

13 _____

14 _____

15 _____

FIGURE 9.4   Anterior view of skull.

5. Hyoid Bone

The hyoid bone is not a part of the skull and is not attached to the axial skeleton, but it is included with the axial skeleton because of its location in the anterior neck region between the mandible and the larynx. This U-shaped bone has the distinction of not articulating with any other bones. It is secured in place by ligaments and muscles, including many muscles of the tongue and neck. This is the bone that is crushed during strangulation.

6. Paranasal Sinuses, Hard Palate, and Nasal Septum

Paranasal (*para-* = next to; *nasal* = nose) **sinuses** (*sinu-* = hollow) are cavities lined with mucous membranes that are located near and have openings into the nasal cavities. The four paranasal sinuses—the **ethmoidal, frontal, maxillary,** and **sphenoidal**—are named for their locations in bones of the same name.

The **hard palate,** or roof of the mouth, is formed by the fusion of four bones; two **maxillary bones** and two **palatine bones.** Approximately three-fourths of the hard palate is composed of the fused maxillary bones and one-fourth is the fused palatine bones.

The **nasal septum** (*saeptum* = wall or partition) is comprised of bone and cartilage. The inferior bone is the **vomer,** and the **perpendicular plate** of the **ethmoid bone** is the superior bone. Cartilage is anterior to the two bones.

CLINICAL NOTE: *A **cleft palate** is a common congenital malformation of the hard palate. This defect is the result of bones of the palate failing to fuse along the midline during embryonic development. The cleft can be partial or complete.*

ACTIVITY 2 HYOID BONE, PARANASAL SINUSES, HARD PALATE, AND NASAL SEPTUM

1 Locate the hyoid bone inferior to the mandible in Figure 9.1.

2 Locate the bones forming the hard palate in Figure 9.2.

3 Locate the bones forming the nasal septum in Figure 9.4.

4 Identify the hyoid bone on an articulated skeleton and with a disarticulated skeleton.

5 Locate the bones forming the hard palate and nasal septum on a skull.

6 Palpate your hyoid bone by placing the thumb and middle finger of one hand on either side of the neck about one inch inferior to the mandible. This bone can be moved laterally from side to side.

7 Label and color the four paranasal sinuses in Figure 9.5.

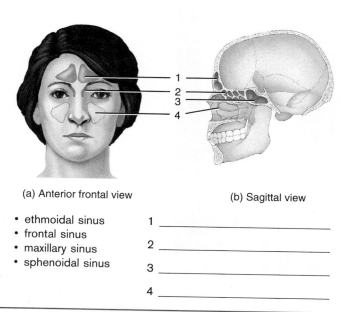

(a) Anterior frontal view   (b) Sagittal view

- ethmoidal sinus
- frontal sinus
- maxillary sinus
- sphenoidal sinus

1 _____

2 _____

3 _____

4 _____

FIGURE 9.5   Paranasal sinuses.

7. Fetal Skull

The fetal skull is not completely ossified, but has membranous sections composed of fibrous connective tissue called **fontanels** (*fontaine* = fountain) or "soft spots." Fontanels allow the cranial bones to compress during the journey through the birth canal, and also allow for future growth of the brain. At birth, the physician or midwife can tell how the baby is presenting itself in the birth canal by palpating the fontanel. If the fontanel is diamond shaped, the baby is presenting the frontal bone first and is face down. If the fontanel is triangular in shape, the baby is presenting the occipital bone and is face up. These membranous templates that are present at birth eventually ossify.

ACTIVITY 3 FETAL SKULL AND MAJOR FONTANELS

1 Label the fontanels in Figure 9.6(a) and (b).

2 Identify the fontanels on a fetal skull (if available).

3 Compare the location of the fontanels and sutures in the fetal skull with adult sutures.

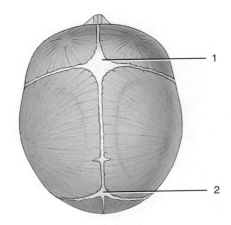

(a) Superior view

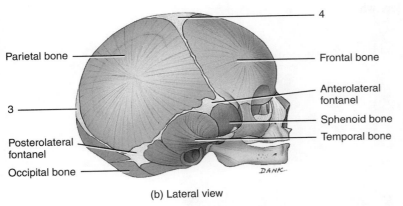

Parietal bone

Frontal bone

Anterolateral fontanel

Sphenoid bone

Temporal bone

3

Posterolateral fontanel

Occipital bone

DANK

(b) Lateral view

(a) Superior view
• anterior (frontal) fontanel
• posterior (occipital) fontanel

1 _____

2 _____

(b) Lateral view
• anterior (frontal) fontanel (fon-ta-NEL)
• posterior (occipital) fontanel

3 _____

4 _____

FIGURE 9.6   Major fontanels of the fetal skull.

C. VERTEBRAL COLUMN

The curved **vertebral column** (backbone or spine) protects the spinal cord and provides attachment points for back muscles. This structure is flexible and can be bent, twisted, and rotated, especially in the cervical region. The vertebral column consists of **cervical** (*cervic-* = neck), **thoracic** (*thorac-* = chest), and **lumbar** (*lumb-* = loin) **vertebrae** in addition to the **sacrum** (*sacer* = sacred) and **coccyx** (*coccyx* = cuckoo's beak). The sacrum consists of five vertebrae that are fused in the adult. The coccyx (tailbone) is usually composed of four small, fused vertebrae. An infant has 33 individual vertebrae and the adult has 26 due to fusion of the sacral and coccygeal vertebrae. The vertebral column articulates with the skull superiorly, the ribs laterally, and the pelvis inferiorly.

1. Regions and Normal Curvatures of the Vertebral Column

The five regions of the vertebral column and number of vertebrae in the adult from superior to inferior, are: **cervical** (7), **thoracic** (12), **lumbar** (5), **sacral** (1), and **coccygeal** (1). To remember the number of vertebrae in the first three groups, students say they eat "breakfast at 7, lunch at 12, and dinner at 5." There are four normal spinal curvatures that correspond with the regions of the spinal column: **cervical, thoracic, lumbar,** and **sacral** (pelvic) **curvatures.** A newborn has a single primary spinal curve that will become the thoracic curve and sacral curve. Two secondary curves, the cervical and lumbar, develop several months later. The cervical curve develops when the baby can hold its head erect and the lumbar curve develops when the baby can stand.

ACTIVITY 4 THE VERTEBRAL COLUMN

1 Label the five regions of the vertebral column in Figure 9.7.

2 Label the four normal spinal curvatures in Figure 9.7.

3 Identify the five regions on an articulated skeleton or a vertebral column model.

4 Identify the four spinal curvatures on an articulated skeleton or vertebral column.

5 Pronounce the terms as you point to them on a skeleton or vertebral column.

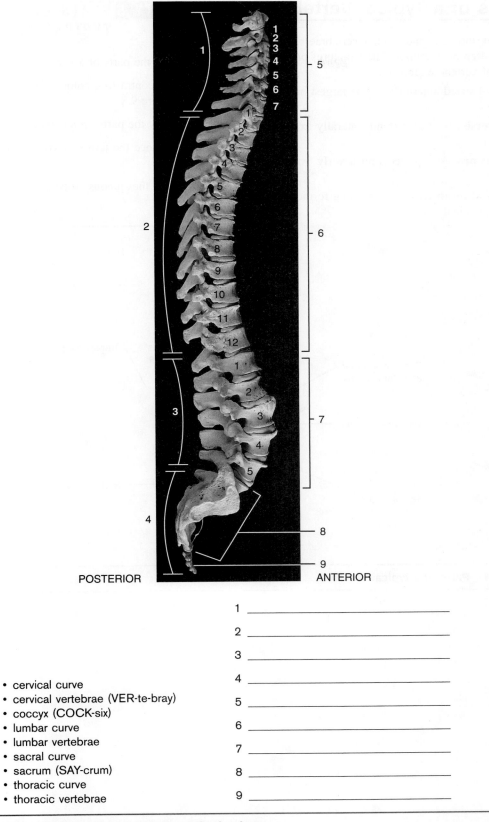

POSTERIOR   ANTERIOR

1 _____

2 _____

3 _____

4 _____

- cervical curve   5 _____
- cervical vertebrae (VER-te-bray)
- coccyx (COCK-six)   6 _____
- lumbar curve
- lumbar vertebrae   7 _____
- sacral curve
- sacrum (SAY-crum)   8 _____
- thoracic curve
- thoracic vertebrae   9 _____

FIGURE 9.7   Normal spinal curvatures of the vertebral column.

2. Parts of a Typical Vertebra

The cervical, thoracic, and lumbar vertebrae have similar as well as different structures. The structures present in all three type of vertebrae are:

- **body**—located anteriorly; is the largest part of the vertebra

- **transverse process**—extends laterally from the vertebra

- **spinous process**—projects posteriorly from the vertebra

- **vertebral foramen**—large opening formed that protects the spinal cord

ACTIVITY 5 PARTS OF A TYPICAL VERTEBRA

1 Locate the parts of a vertebra shown in Figure 9.8.

2 Using contrasting colors, color the parts of a vertebra in Figure 9.8.

3 Identify the parts on a disarticulated thoracic vertebra.

4 Pronounce the terms as you point to them on the vertebra.

5 Palpate the spinous processes of your vertebral column.

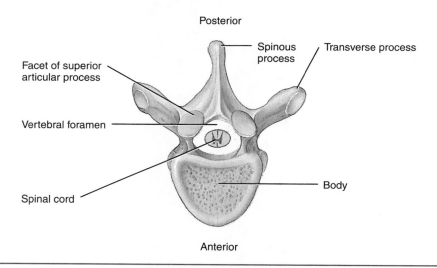

FIGURE 9.8   **Parts of a typical vertebra.**

3. Comparison of Cervical, Thoracic, and Lumbar Vertebrae

The first two of the seven cervical vertebrae look different from the others. The **atlas** is C1 and is named for the mythical Greek god who held up the world on his shoulders. The large **superior articular facets** of the atlas articulate with the **occipital condyles** of the skull and allow the head to move up and down in a nodding motion. The **axis,** C2, has a superior tooth-like protuberance called the **dens** (*dens* = tooth in Latin) or **odontoid process** (*odous* = tooth in Greek). The dens extends superiorly into the vertebral foramen of the atlas and allows the atlas to pivot laterally in a "no" type motion. Some head traumas push the head onto the vertebral column and the dens is shoved into the medulla oblongata (brain stem). This is the cause of death in such head injuries. The easiest way to distinguish any **cervical vertebra** is by its three foramina (plural)—a **vertebral foramen** plus two **transverse foramina.** The transverse foramina, found only in cervical vertebrae, allow passage of the vertebral arteries, veins, and sympathetic nerves to and from the brain. Since the cervical vertebrae do not carry much weight, they are lightweight and have small bodies. C3 through C6 have bifurcated (forked) spinous processes. A **whiplash,** which occurs in some rear-ended auto accidents, causes partial or complete dislocation of the cervical vertebrae.

A **thoracic vertebra** has a medium-sized body and a long, narrow spinous process that commonly slants inferiorly at a sharp angle. Thoracic vertebrae also have facets (*facette* = little face) on the transverse processes and demifacets (*demi-* = half) on the bodies, both of which articulate with ribs.

Lumbar vertebrae have the largest bodies to support more weight, and thick, hatchet-shaped spinous processes that extend horizontally.

Intervertebral discs can be found between the bodies of vertebrae from C2 to the sacrum. They are made of fibrocartilage to give strength, permit movement, and absorb shock. A *herniated* or *slipped disc* occurs when the fibrocartilage is stressed and torn or cracked. In this situation, the inner pulp-like center protrudes outward and causes pressure on the spinal cord or a spinal nerve.

Intervertebral foramina are formed when vertebrae are stacked on one another. Spinal nerves exit the vertebral column through these foramina.

ACTIVITY 6 COMPARISON OF CERVICAL, THORACIC, AND LUMBAR VERTEBRAE

1 Identify the specific parts of an atlas, axis, and cervical vertebra in Figure 9.9(a), (b), and (c).

2 Identify the specific parts of a thoracic vertebra in Figure 9.10(a) and (b) and a lumbar vertebra in Figure 9.11(a) and (b).

3 Pronounce each term as you point to it.

4 Using disarticulated vertebrae, place all three types side by side and note their differences. Mix them up and identify each type of vertebra with a partner.

5 Identify the parts of articulated vertebrae in Figure 9.12(a) and a herniated disc in Figure 9.12(b).

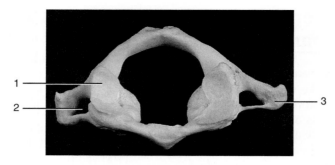

(a) Superior view of the atlas (C1)

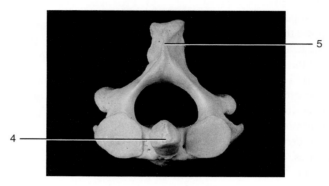

(b) Superior view of the axis (C2)

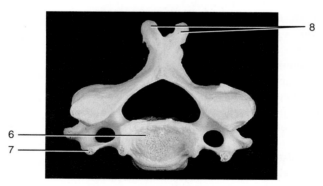

(c) Superior view of a typical cervical vertebra (C3)

(a) Atlas
• superior articular facet
• transverse foramen
• transverse process

(b) Axis (Note: The transverse foramen cannot be seen in this photo.)
• dens or odontoid process
• spinous process

1 _____

2 _____

3 _____

4 _____

5 _____

(c) Typical cervical vertebra
• bifurcated spinous process
• body
• transverse process

6 _____

7 _____

8 _____

FIGURE 9.9 Cervical vertebrae.

POSTERIOR

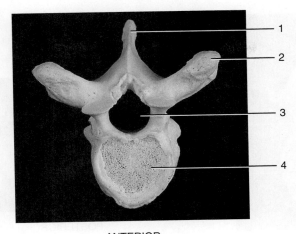

ANTERIOR

(a) Superior view

SUPERIOR

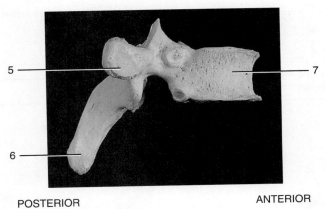

POSTERIOR   ANTERIOR

(b) Right lateral view

(a) Superior view
• body
• spinous process
• transverse process
• vertebral foramen

(b) Right lateral view
• body
• facet on transverse process
• slanted spinous process

1 _____

2 _____

3 _____

4 _____

5 _____

6 _____

7 _____

FIGURE 9.10   **Thoracic vertebrae.**

POSTERIOR

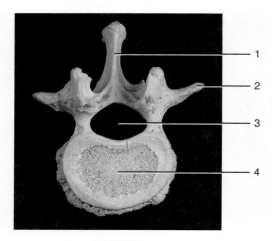

ANTERIOR

(a) Superior view

SUPERIOR

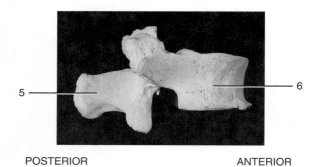

POSTERIOR   ANTERIOR

(b) Right lateral view

(a) Superior view
• body
• spinous process
• transverse process
• vertebral foramen

1 _____

2 _____

3 _____

4 _____

(b) Right lateral view
• body
• hatchet-shaped spinous process

5 _____

6 _____

FIGURE 9.11 **Lumbar vertebrae.**

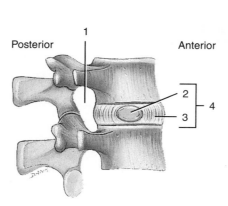

(a) Right lateral view of articulated vertebrae
with intervertebral disc

Posterior

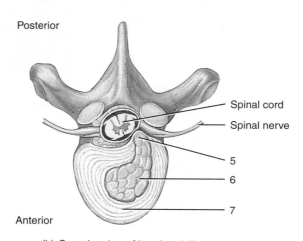

Spinal cord

Spinal nerve

Anterior

(b) Superior view of herniated disc

(a) Articulated vertebrae
• inner pulp-like center
• intervertebral disc
• intervertebral foramen
• outer fibrous part

1 _____

2 _____

3 _____

4 _____

(b) Herniated disc
• herniation
• inner pulp-like center
• outer fibrous part

5 _____

6 _____

7 _____

FIGURE 9.12 **Intervertebral discs.**

4. Sacrum and Coccyx

The **sacrum** has a slightly curved, triangular shape and has two **superior articular processes** with facets that articulate with the fifth lumbar vertebra. The sacrum has **sacral foramina** that provide exits for spinal nerves. From the posterior view, the opening of the **sacral canal** is located just posterior to the body of the sacrum and is open all the way through the sacrum. (*Note:* On many plastic models, the sacral canal is closed.) The lateral surfaces of the sacrum, the **auricular surfaces,** are roughened to articulate with the iliac portion of the os coxa on each side, forming **sacroiliac joints.** The vertebral column ends with the small **coccyx,** or tailbone. The tiny, fused vertebrae of the coccyx are attached to the sacrum with ligaments.

ACTIVITY 7 SACRUM AND COCCYX

1 Locate the parts of the sacrum and coccyx in Figure 9.13.

2 Identify these structures on a model of the sacrum and coccyx.

3 Pronounce the terms as you point to them on the model.

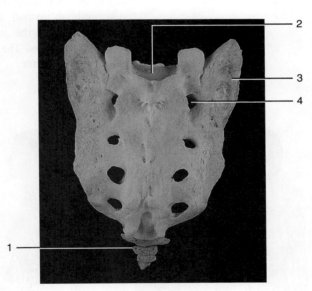

Posterior view

- auricular surface (for sacroiliac joint)
- coccyx
- sacral canal
- sacral foramen

1 _____

2 _____

3 _____

4 _____

FIGURE 9.13   Sacrum and coccyx.

5. Abnormal Spinal Curvatures

Three abnormal curves of the vertebral column are scoliosis, kyphosis, and lordosis. The vertebral column bends laterally in **scoliosis** (*scolio-* = crooked). **Kyphosis** is an exaggerated thoracic curve (kyphotic curvature) that results in a hunched back with rounded shoulders. **Lordosis** is an exaggerated lumbar curve (lordotic curvature) that appears as a swayback with the abdomen protruding anteriorly.

ACTIVITY 8 ABNORMAL CURVES OF THE VERTEBRAL COLUMN

Identify the three different types of abnormal curves in Figure 9.14(a), (b), and (c).

D. THORACIC CAGE

The bony cage that encircles the chest is called the **thoracic cage.** The thoracic cage is composed of the sternum, ribs, costal cartilages, and bodies of the thoracic vertebrae.

The **sternum** is a narrow flat bone that is composed of three fused bones: the manubrium, the body of the sternum, and the xiphoid process. The **manubrium** (*manubrium* = handle) is the portion of the sternum superior to the **body.** The manubrium and body articulate (*articul-* = pertaining to a joint) with the **costal cartilages** of the **ribs.** The **xiphoid process** (*xiphoid* = sword-like) is the inferior portion of the sternum that is shaped like a small sword.

There are 12 pairs of **ribs** in both males and females. The first seven pairs are called the **true ribs** because their **costal (rib) cartilages** have a direct attachment to the sternum. The last five rib pairs (8 through 12) are called **false ribs.** Rib pairs 8 through 10 do not have a direct attachment to the sternum, but attach to the costal cartilage of the seventh rib instead. Rib pairs 11 and 12 do not have any attachment to the sternum or cartilage and are also called **floating ribs.** The space between the ribs is called the **intercostal space.**

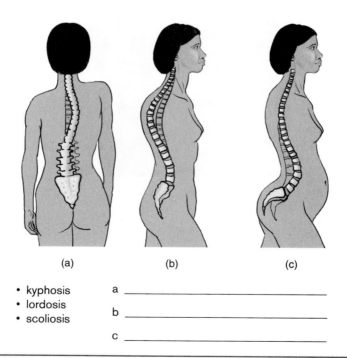

(a)   (b)   (c)

- kyphosis
- lordosis
- scoliosis

a _____

b _____

c _____

FIGURE 9.14   Abnormal curves of the vertebral column.

ACTIVITY 9 THORACIC CAGE AND RIB ARTICULATIONS

1 Locate the parts of the thoracic cage in Figure 9.15.

2 Identify these parts on an articulated thoracic cage or skeleton.

3 Pronounce the name of each part as you point to it.

4 Palpate as many of these parts as you can on your body.

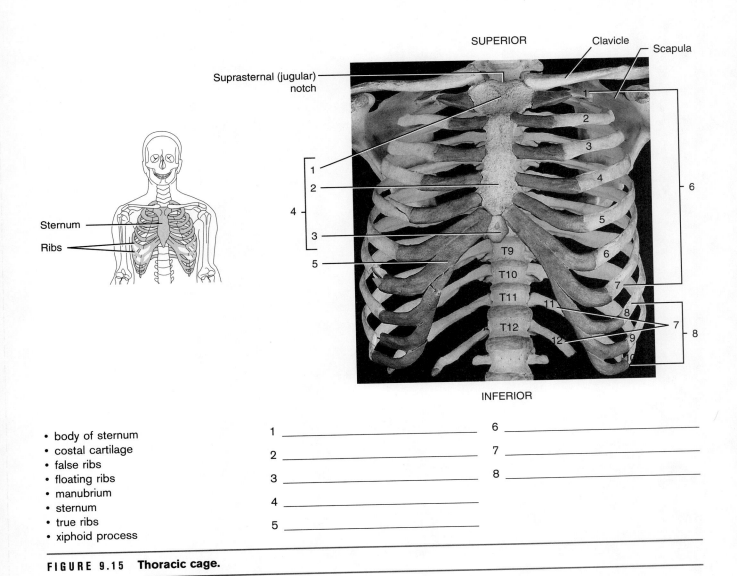

- body of sternum
- costal cartilage
- false ribs
- floating ribs
- manubrium
- sternum
- true ribs
- xiphoid process

1 _____   6 _____

2 _____   7 _____

3 _____   8 _____

4 _____

5 _____

FIGURE 9.15   Thoracic cage.

REVIEWING YOUR KNOWLEDGE

A. Vertebral Column and Thoracic Cage

Match the skeletal term with the proper description.

_____ **1.** Soften jolts to the vertebral column

_____ **2.** Opening for spinal nerve exit

_____ **3.** Bony protection for thoracic organs

_____ **4.** First cervical vertebra

_____ **5.** Second cervical vertebra

_____ **6.** Articulates with rib anteriorly

_____ **7.** Tailbone

_____ **8.** Encloses and protects the spinal cord

_____ **9.** Lordotic curvatures

_____ **10.** Kyphotic curvatures

a. atlas

b. axis

c. cervical and lumbar curvatures

d. coccyx

e. intervertebral discs

f. intervertebral foramen

g. sternum

h. thoracic cage

i. thoracic and pelvic curvatures

j. vertebral column

B. Other Major Features of the Axial Skeleton

Write in the major skull feature or bone that is described.

_____   **1.** Atlas articulates superiorly with this bone

_____   **2.** Facial bone that contains a paranasal sinus

_____   **3.** Articulation between the mandible and skull

_____   **4.** Neck and tongue muscles attach to this bone

_____   **5.** Structures that allow the baby's cranium to compress during childbirth

_____   **6.** Special feature present on the transverse process of cervical vertebrae

_____   **7.** Bone that has the odontoid process

_____   **8.** Sacrum articulates superiorly with this bone

_____   **9.** Inferior bone of the nasal septum

_____   **10.** Bone marking that protects the pituitary gland

C. Bones of the Axial Skeleton

Fill in the blank with the correct bone.

_____   **1.** Middle and inner ear are found within this bone

_____   **2.**
                                    } Contain teeth sockets
_____   **3.**

_____   **4.** Bone that does not articulate with any another bone

_____   **5.** Only freely movable skull bone

_____   **6.** Ribs articulate posteriorly with these bones

_____   **7.** Chin bone

_____   **8.** Smaller bone of the hard palate

_____   **9.** Upper jaw bone

_____   **10.** Cheek bones

_____   **11.** Tear ducts pass through these paired bones

APPENDICULAR SKELETON

10

OBJECTIVES

After completing this exercise, you should be able to:

- Identify the principal bone markings of the pectoral (shoulder) girdle and the upper limb
- Describe the articulation of the pectoral girdle with the humerus
- Describe the articulation of the humerus with the forearm bones
- Identify the bone markings of the pelvic girdle and the lower limbs
- Describe the articulation of the bones of the pelvic girdle with the femur
- Describe the articulation of the femur with the leg bones

MATERIALS

- articulated skeleton
- disarticulated skeleton
- colored pencils

The **appendicular skeleton** (*appendere* = to add something) has larger bones than the axial skeleton and bears more weight. This division of the skeleton is composed of 126 bones that make up the upper and lower limbs or extremities, and the bones or girdles that attach the limbs to the axial skeleton. The bones of this division are separated into four main areas: the **pectoral girdles**

(*pectus* = breast; *girdle* = any curved structure), the **upper limbs,** the **pelvic girdle** (*pelvis* = basin), and the **lower limbs.** The bone markings include sites of muscle attachment and articulations with other bones to form a joint. Bone surface markings that pertain to the appendicular skeleton are given in Table 10.1.

TABLE 10.1 Selected Bone Surface Markings of the Appendicular Skeleton

MARKING	DESCRIPTION	PURPOSE
1. Crest	Prominent ridge	Muscle attachment
2. Condyle	Smooth, rounded articular process	Articulation
3. Epicondyle	Projection above a condyle	Muscle attachment
4. Foramen	Opening or hole	Opening for blood vessels and nerves
5. Fossa	Shallow depression	Muscle attachment or articulation
6. Head	Rounded articular projection supported on the neck of a bone	Articulation
7. Notch	An indentation or depression	Groove for blood vessels and nerves
8. Trochanter	Very large projection	Muscle attachment
9. Spine	Sharp, slender projection	Muscle attachment
10. Tuberosity	Large, roughened projection	Muscle attachment

A. PECTORAL GIRDLE

There are two pectoral girdles and each attaches an upper limb to the axial skeleton. Each **pectoral** (or shoulder) **girdle** is composed of a **scapula** (shoulder blade) and a **clavicle** (*clavicle* = key).

The lateral end of the clavicle articulates with the acromion (acromial process) of the scapula. These two bones form the pectoral girdle. The medial end of the clavicle articulates with the manubrium of the sternum, attaching the pectoral girdle to the axial skeleton. The scapula does not articulate directly with the axial skeleton, but is attached to it with muscles.

1. Bones and Selected Bone Markings of the Pectoral Girdle

Clavicle (Collar bone)

Scapula (Shoulder blade)

- **spine**—sharp ridge located on posterior side
- **acromion**—flattened process at lateral end of spine that articulates with the clavicle
- **glenoid cavity** (*glene* = joint socket) or **fossa**—depression inferior to acromion that articulates with the head of the humerus
- **coracoid process** (*coracoid* = crow's beak)—superior and medial to glenoid cavity; site for muscle attachment

ACTIVITY 1 BONES AND BONE MARKINGS OF THE PECTORAL GIRDLE

1 Identify the selected bone markings of the clavicle and scapula in Figure 10.1(a), (b), and (c).

2 Locate these structures on an articulated skeleton and on disarticulated bones.

3 Pronounce the terms as you point to them.

4 Palpate these structures on your own body: clavicle, acromion (process), and spine of scapula.

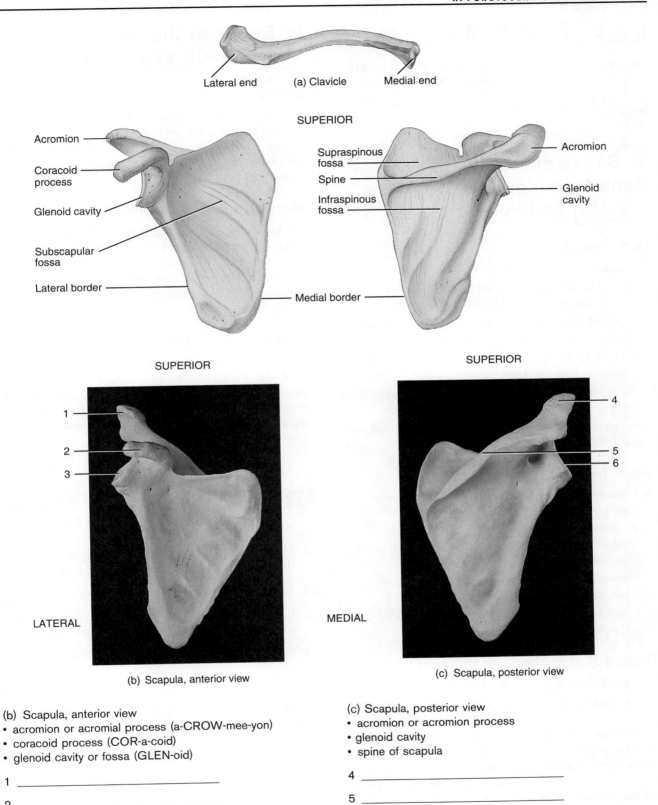

Lateral end   (a) Clavicle   Medial end

SUPERIOR

Acromion
Coracoid process
Glenoid cavity
Subscapular fossa
Lateral border

Supraspinous fossa
Spine
Infraspinous fossa
Acromion
Glenoid cavity

Medial border

SUPERIOR

1
2
3

LATERAL

(b) Scapula, anterior view

SUPERIOR

4
5
6

MEDIAL

(c) Scapula, posterior view

(b) Scapula, anterior view
• acromion or acromial process (a-CROW-mee-yon)
• coracoid process (COR-a-coid)
• glenoid cavity or fossa (GLEN-oid)

1 _____
2 _____
3 _____

(c) Scapula, posterior view
• acromion or acromion process
• glenoid cavity
• spine of scapula

4 _____
5 _____
6 _____

FIGURE 10.1   **The right pectoral girdle.**

B. UPPER LIMB

The upper limb consists of the **humerus, ulna, radius, carpals, metacarpals,** and **phalanges.** Of the 30 bones in each upper limb, one is in the arm, two are in the forearm, and the other 27 are in the hand.

1. Bones and Selected Bone Markings of the Arm and Forearm

Humerus (Arm bone)

- **head**—rounded, proximal end
- **deltoid tuberosity**—raised area between the proximal and distal ends of the humerus
- **trochlea** (*trochlea* = pulley)—spool-shaped medial condyle on the anterior, distal end
- **capitulum** (*caput* = head)—rounded, knob-like condyle lateral to trochlea
- **coronoid fossa** (*corona* = crown)—shallow anterior depression on distal end
- **olecranon fossa** (*olekranon* = tip of the elbow)—largest depression on posterior, distal end

Ulna (Medial bone of forearm)

- **olecranon**—large, curved lip-like projection on posterior side of proximal end
- **coronoid process**—smaller, curved lip-like projection on anterior side of proximal end; distal to olecranon
- **trochlear notch**—deep, curved area between olecranon and coronoid process
- **styloid process** (*stylos* = pole; *-oid* = like)—slender, pointed projection on distal end

Radius (Lateral bone of forearm)

- **head**—flat, disc-shaped proximal end
- **styloid process**—slender, pointed projection; distal end

The shoulder joint that connects the upper limb to the pectoral girdle is formed by the head of the humerus (*humeri* = shoulder) articulating with the glenoid cavity of the scapula. The coronoid process and olecranon of the ulna fit into the coronoid and olecranon fossae (plural) of the humerus. The trochlea of the humerus articulates with the trochlear notch of the ulna. The head of the radius articulates with the capitulum of the humerus at the elbow joint and with the radial notch of the ulna at the proximal radioulnar joint.

2. Bones of the Hand: Carpals, Metacarpals, and Phalanges

Carpus (Wrists)

The carpus is composed of eight short bones of the wrist, the **carpal** bones, that are lined up to form a proximal and a distal row of bones. Two of the carpal bones articulate with the radius, but there is no articulation of the carpal bones with the ulna.

Metacarpus (Palm of hand)

The metacarpus is composed of five **metacarpal** bones that make up the palm of the hand. They are numbered as Roman numerals I to V (or 1 to 5) from the metacarpal of the thumb (lateral side) to the little finger side. The metacarpals articulate with the carpals proximally and with the phalanges distally.

Phalanges (Fingers)

The phalanges (singular = phalanx) make up the fingers or digits. They are also numbered I to V (or 1 to 5) from the thumb to the little finger. The thumb has two phalanges, proximal and distal. Each of the other four digits has **proximal, middle,** and **distal phalanges** (*phalanges* = line of soldiers).

ACTIVITY 2 THE UPPER LIMB BONES AND BONE MARKINGS

1 Identify the parts of the humerus on Figure 10.2(a) and (b), and the ulna and radius on Figure 10.3(a) and (b).

2 Identify the parts of the hand on Figure 10.4.

3 Locate these bones and their bone markings on an articulated skeleton and on disarticulated bones.

4 Pronounce the terms as you point to them.

5 Palpate the following bone parts on your own body: head of the humerus, olecranon and styloid process of the ulna, and the styloid process of the radius.

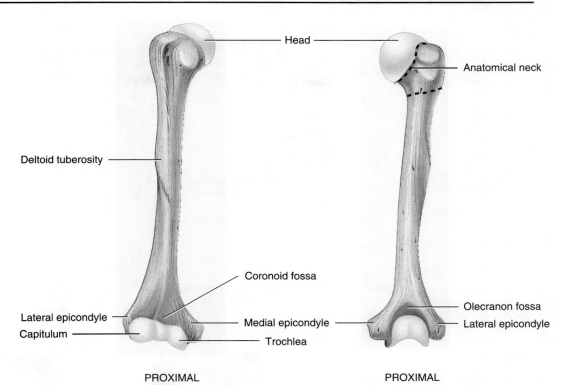

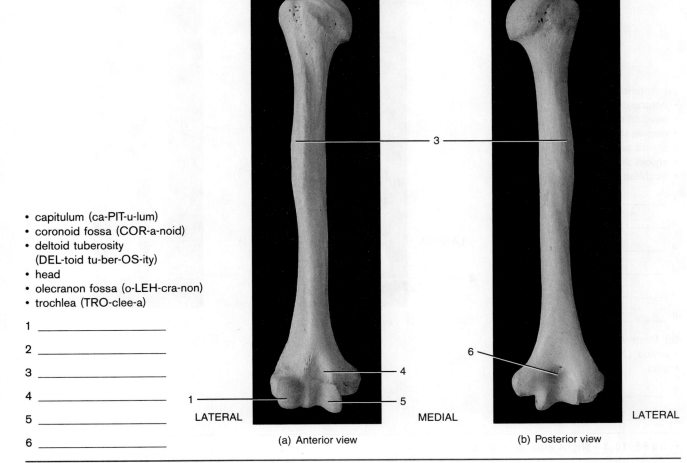

- capitulum (ca-PIT-u-lum)
- coronoid fossa (COR-a-noid)
- deltoid tuberosity
  (DEL-toid tu-ber-OS-ity)
- head
- olecranon fossa (o-LEH-cra-non)
- trochlea (TRO-clee-a)

1 _____

2 _____

3 _____

4 _____

5 _____

6 _____

FIGURE 10.2   **Right humerus.**

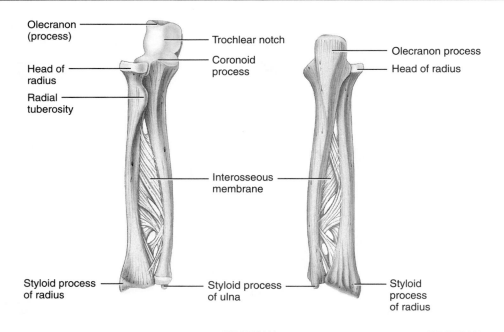

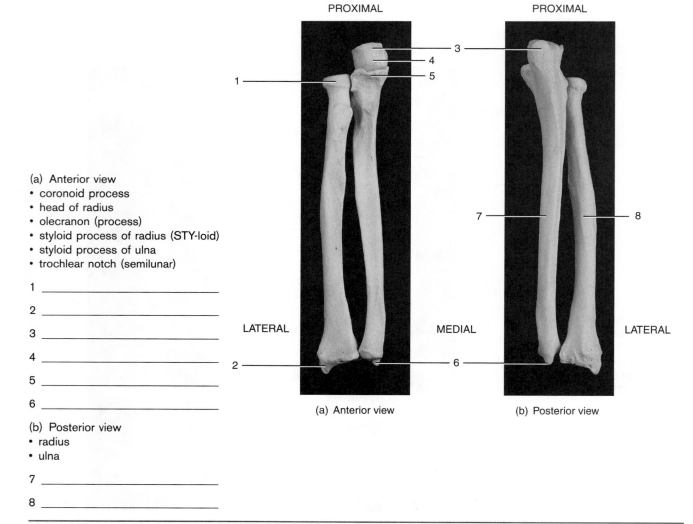

(a) Anterior view
- coronoid process
- head of radius
- olecranon (process)
- styloid process of radius (STY-loid)
- styloid process of ulna
- trochlear notch (semilunar)

1 _____

2 _____

3 _____

4 _____

5 _____

6 _____

(b) Posterior view
- radius
- ulna

7 _____

8 _____

FIGURE 10.3 **Right ulna and radius.**

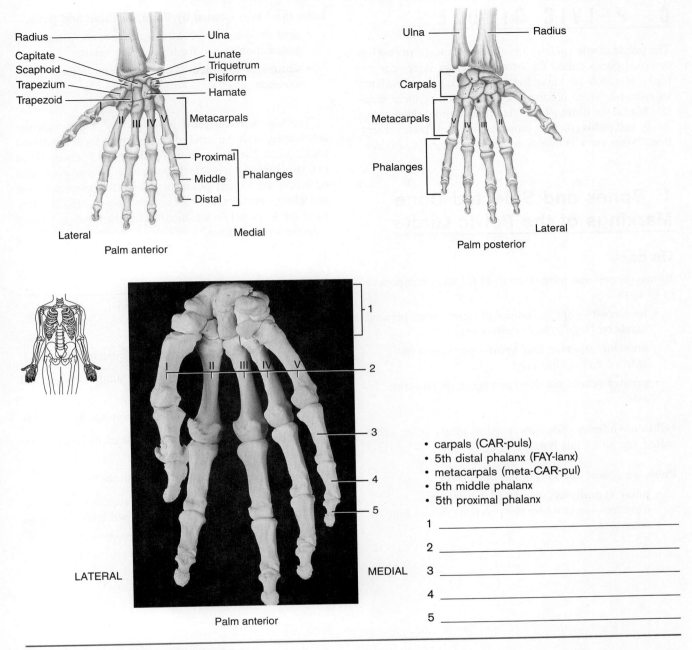

Radius
Ulna
Capitate
Scaphoid
Trapezium
Trapezoid
Lunate
Triquetrum
Pisiform
Hamate
Metacarpals
Proximal
Middle
Distal
Phalanges
Lateral
Medial
Palm anterior

Ulna
Radius
Carpals
Metacarpals
Phalanges
Lateral
Palm posterior

LATERAL
MEDIAL
Palm anterior

- carpals (CAR-puls)
- 5th distal phalanx (FAY-lanx)
- metacarpals (meta-CAR-pul)
- 5th middle phalanx
- 5th proximal phalanx

1 _____

2 _____

3 _____

4 _____

5 _____

FIGURE 10.4   Right hand and wrist.

C. PELVIC GIRDLE

The **pelvic girdle** (*pelvis* = basin) is composed of two hip or coxal bones called the **os coxae** (*os* = bone; *cox-* = hip) that attach the lower limb to the axial skeleton. Each **os coxa** (singular) is formed by the fusion of three separate bones: the **ilium** (*ilia* = flank), **ischium** (*ischion* = hip joint), and **pubis** (*pub-* = grown-up or adult) bones. These three bones are identifiable as separate bones in children.

1. Bones and Selected Bone Markings of the Pelvic Girdle

Os coxa

Ilium—largest and most superior of the three components of os coxa

- **iliac crest**—superior border of ilium; when placing hands on "hips" you're on this area

- **anterior superior iliac spine**—protrusion on anterior end of iliac crest

- **greater sciatic notch**—large notch on posterior side

Ischium—inferior, posterior portion of os coxa; when seated you sit on this bone

Pubis—anterior, inferior portion of os coxa

- **pubic symphysis** (*symphysis* = growing together)—joint where the two pubic bones join anteriorly

Bone markings formed by ilium, ischium, and pubis

- **acetabulum** (*acetabulum* = little saucer)—deep indentation or cup for head of the femur

- **obturator foramen**—largest foramen in the skeleton

The **hip** or **coxal joint** is formed by the acetabulum articulating with the head of the femur to form a ball and socket joint. There is a strong ligament that connects these two structures deep inside the joint itself. When articulated os coxae are united posteriorly with the sacrum at the **sacroiliac joints** (*sacro-* = sacrum; *iliac* = ilium), they form the bony **pelvis** which connects the lower extremity with the axial skeleton.

ACTIVITY 3 THE OS COXAE AND PELVIS

1 Identify the bone markings of an os coxa in Figure 10.5.

2 Using contrasting colors, color the ilium, ischium, and pubis bones in Figure 10.5.

3 Examine the bone markings of the pelvis in Figure 10.6.

4 Locate the markings on a disarticulated os coxa and on an articulated pelvis.

5 Pronounce the terms as you point to them.

6 Palpate the iliac crest, the anterior superior iliac spine, and the pubic symphysis on your own body.

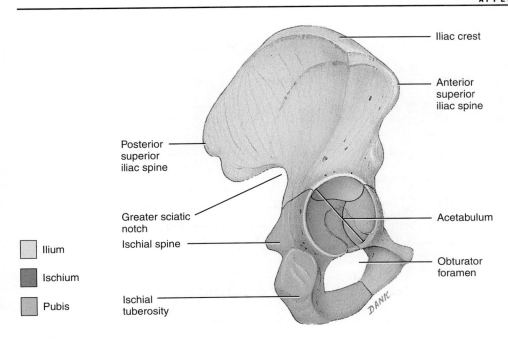

Iliac crest

Anterior superior iliac spine

Posterior superior iliac spine

Greater sciatic notch

Acetabulum

Ischial spine

Obturator foramen

Ischial tuberosity

Ilium

Ischium

Pubis

SUPERIOR

POSTERIOR

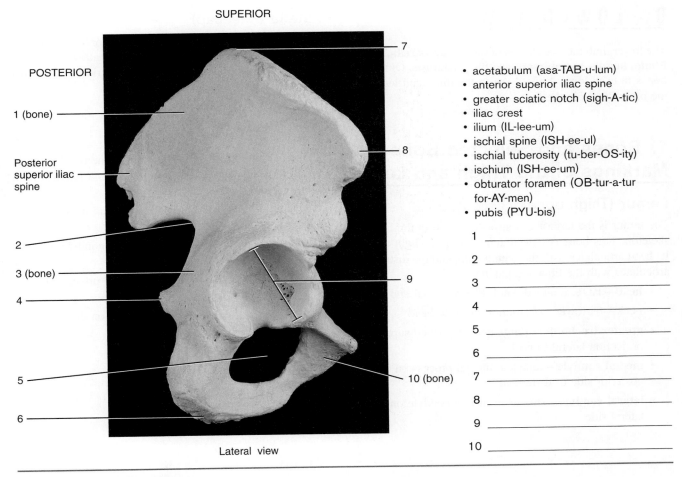

1 (bone)

Posterior superior iliac spine

2

3 (bone)

4

5

6

7

8

9

10 (bone)

Lateral view

- acetabulum (asa-TAB-u-lum)
- anterior superior iliac spine
- greater sciatic notch (sigh-A-tic)
- iliac crest
- ilium (IL-lee-um)
- ischial spine (ISH-ee-ul)
- ischial tuberosity (tu-ber-OS-ity)
- ischium (ISH-ee-um)
- obturator foramen (OB-tur-a-tur for-AY-men)
- pubis (PYU-bis)

1 _____
2 _____
3 _____
4 _____
5 _____
6 _____
7 _____
8 _____
9 _____
10 _____

FIGURE 10.5   **Right os coxa.**

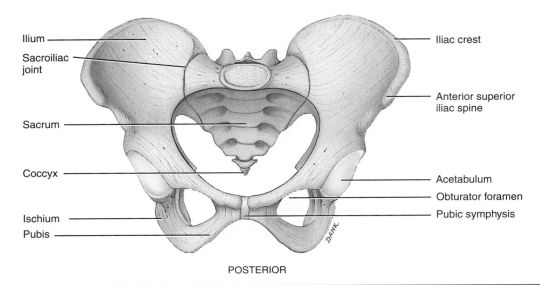

Ilium

Sacroiliac joint

Sacrum

Coccyx

Ischium

Pubis

Iliac crest

Anterior superior iliac spine

Acetabulum

Obturator foramen

Pubic symphysis

POSTERIOR

FIGURE 10.6 Pelvis.

D. LOWER LIMB

The lower limb includes bones of the **femur, patella, tibia, fibula, tarsals, metatarsals,** and the **phalanges.** Of the 30 bones in each lower limb, 4 are in the thigh and leg, and the other 26 are in the foot.

1. Bones and Selected Bone Markings of the Thigh and Leg

Femur (Thigh bone)

The femur is the largest and strongest bone in the human skeleton. This bone is bowed anteriorly in a slight curve. Its head articulates with the acetabulum, and the distal end articulates with the tibia and patella.

* **head**—large, rounded, knob-like proximal end
* **neck**—narrower, constriction distal to head
* **greater trochanter**—large and roughened superior projection lateral to neck
* **medial condyle**—rounded, medial process on posterior side of distal end
* **lateral condyle**—similar to medial condyle on lateral side

Patella (Kneecap)

This small, triangular bone has an anterior surface that is smoother than the posterior surface. Shallow, irregular-shaped articular facets are on the posterior surface that articulate with the condyles of the femur.

Tibia (Leg bone)

The tibia is the weight-bearing bone of the two leg bones and is medially located. Laterally, the tibia forms a joint with the fibula.

* **medial condyle**—flattened, expanded medial projections on proximal end
* **lateral condyle**—similar to medial condyle on lateral side
* **tibial tuberosity**—large, roughened projection on anterior surface, inferior to condyles
* **medial malleolus**—medial process on distal end, forms medial bump of ankle

Fibula (Leg bone)

The slender fibula is the lateral leg bone that is important for muscle attachment, but not for bearing weight. The **head** of the fibula is proximal and articulates with the tibia, but not the femur. The **lateral malleolus** is distal and articulates with the talus laterally.

- **head**—proximal end
- **lateral malleolus**—distal end, forms lateral bump of ankle

2. Bones of the Foot: Tarsals, Metatarsals, and Phalanges

Tarsus (Ankles)

The tarsus is composed of seven tarsal bones of the foot, with two of them being larger than the rest. The largest tarsal bone is the **calcaneus** (*calcaneum* = heel), also known as the heel bone. The other large tarsal bone is the **talus** (*talus* = ankle) that articulates with the tibia.

Metatarsus (Anterior portion of the instep)

The metatarsus is composed of five metatarsal bones (*meta-* = after or next) that coincide with the metacarpals in the hand. They are numbered I to V (or 1 to 5) from the great toe to the little toe.

Phalanges (Toes)

The phalanges (toes or digits) are similar to the phalanges in the hand. They are numbered I to V (or 1 to 5) from the great toe to the little toe. The great toe is made of two phalanges (**proximal** and **distal**) and digits II to V have three bones each, **proximal, middle,** and **distal phalanges.**

ACTIVITY 4 THE LOWER LIMB BONES AND BONE MARKINGS

1 Identify the bones and bone markings of the femur on Figure 10.7(a) and (b), and the patella, tibia, and fibula on Figures 10.8(a) and (b).

2 Label the bones of the foot on Figure 10.9.

3 Locate the bones and bone markings of the thigh and leg on disarticulated bones as well as on an articulated skeleton.

4 Locate the bones of the foot on an articulated foot.

5 Pronounce the terms as you point to them.

6 Palpate these bone markings on your own body: greater trochanter, patella, head of the fibula, tibial tuberosity, medial malleolus, lateral malleolus, calcaneus, and talus.

ACTIVITY 5 ASSEMBLY OF A COMPLETE DISARTICULATED SKELETON

1 Obtain a disarticulated skeleton from your instructor.

2 With your lab partners, assemble the skeleton.

3 Have your instructor check it to see that you have assembled it properly.

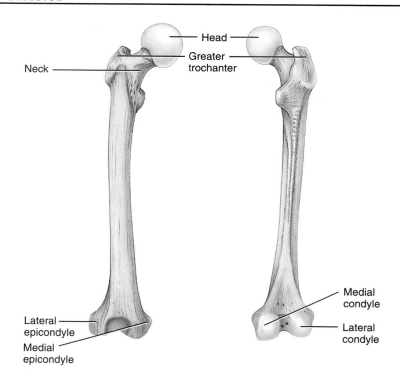

Neck

Head

Greater
trochanter

Lateral
epicondyle

Medial
epicondyle

Medial
condyle

Lateral
condyle

- greater trochanter (tro-CAN-ter)
- head of femur
- lateral condyle (CON-dile)
- medial condyle
- neck

(a) Anterior view

1 _____

2 _____

3 _____

4 _____

(b) Posterior view

5 _____

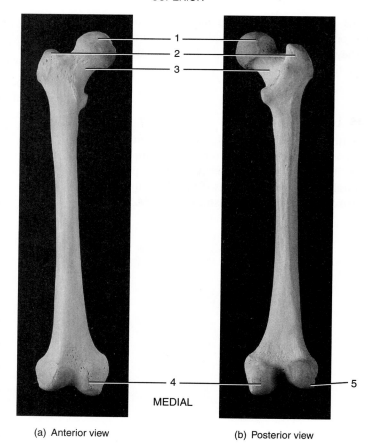

SUPERIOR

1

2

3

4

5

MEDIAL

(a) Anterior view

(b) Posterior view

FIGURE 10.7 Right femur.

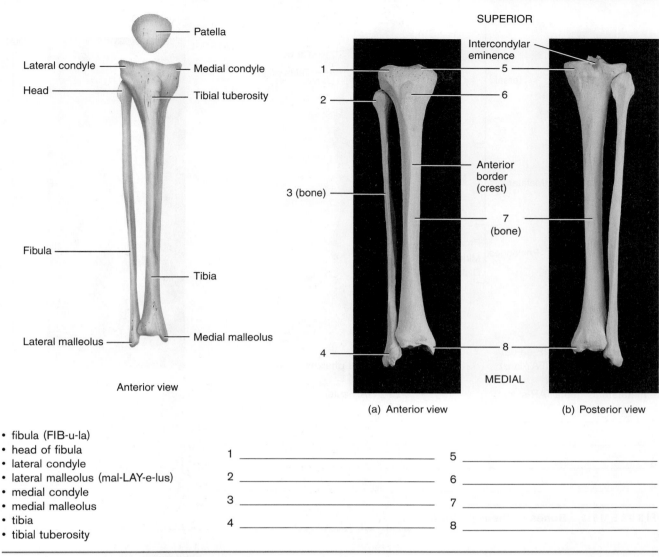

Anterior view

(a) Anterior view   (b) Posterior view

SUPERIOR

MEDIAL

• fibula (FIB-u-la)
• head of fibula
• lateral condyle
• lateral malleolus (mal-LAY-e-lus)
• medial condyle
• medial malleolus
• tibia
• tibial tuberosity

1 _____   5 _____

2 _____   6 _____

3 _____   7 _____

4 _____   8 _____

FIGURE 10.8   **Right tibia, fibula, and patella.**

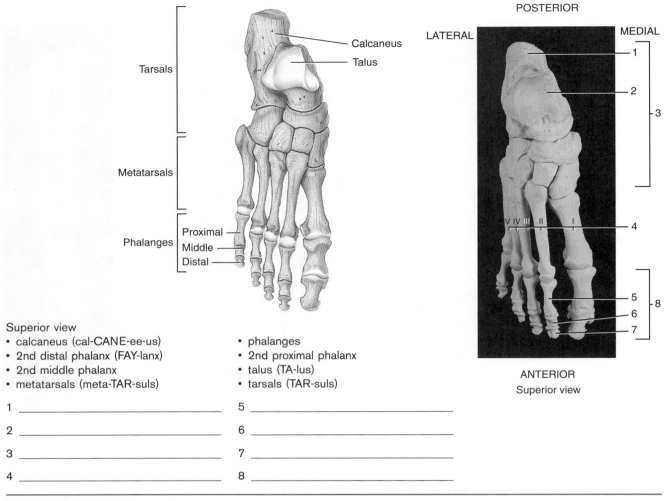

POSTERIOR

LATERAL

MEDIAL

Tarsals

Calcaneus

Talus

Metatarsals

Phalanges

Proximal

Middle

Distal

Superior view
• calcaneus (cal-CANE-ee-us)
• 2nd distal phalanx (FAY-lanx)
• 2nd middle phalanx
• metatarsals (meta-TAR-suls)

• phalanges
• 2nd proximal phalanx
• talus (TA-lus)
• tarsals (TAR-suls)

ANTERIOR
Superior view

1 _____   5 _____

2 _____   6 _____

3 _____   7 _____

4 _____   8 _____

FIGURE 10.9   **Bones of the right foot.**

REVIEWING YOUR KNOWLEDGE

A. Bones and Bone Markings of the Pectoral Girdle and Upper Limb

Identify the bones and bone markings in Figure 10.10.

1. _____

2. _____

3. _____

4. _____

5. _____

6. _____

7. _____

8. _____

9. _____

10. _____

11. _____

12. _____

13. _____

14. _____

15. _____

16. _____

17. _____

18. _____

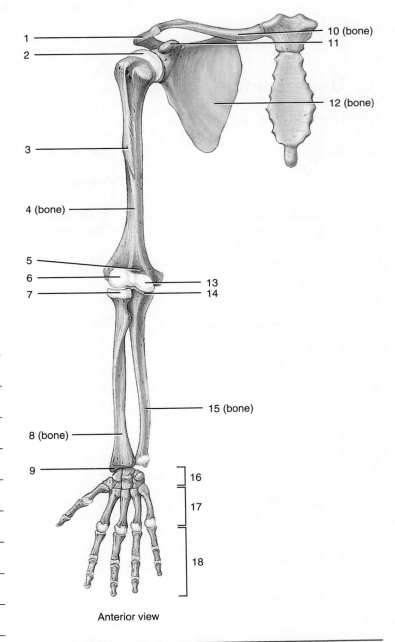

Anterior view

FIGURE 10.10   Pectoral girdle and upper limb.

B. Pectoral Girdle (Shoulder)

Fill in the blank with the correct term.

_____ 1. The acromion (process) articulates with what bone?

_____ 2. Is the clavicle anterior or posterior compared to the scapula?

_____ 3. The clavicle articulates with which bone medially?

_____ 4. The humerus articulates with what bone marking of the scapula?

_____ 5. Name the larger bone that makes up the pectoral girdle.

_____ 6. Name the smaller bone that makes up the pectoral girdle.

C. Upper Limb

Fill in the blank with the correct term.

_____ 1. What part of the radius articulates with the humerus?

_____ 2. What part of the ulna fits into the olecranon fossa of the humerus?

_____ 3. The coronoid process articulates with what depression on the distal end of the humerus?

_____ 4. Is the ulna medial or lateral compared with the radius?

_____ 5. What are the bones called that make up the palm of the hand?

_____ 6. What are the bones of the wrist called?

D. Bones and Bone Markings of the Pelvic Girdle and the Lower Limb

Identify the bone and bone markings in Figure 10.11.

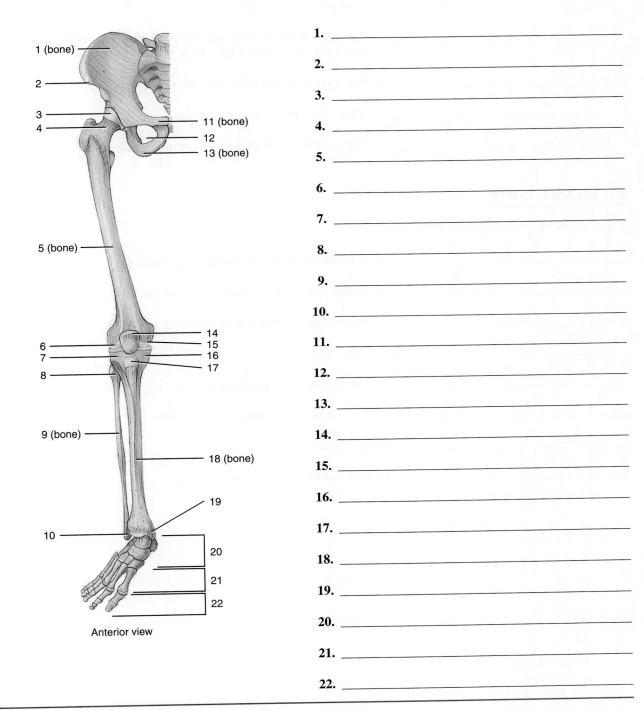

Anterior view

1. _____

2. _____

3. _____

4. _____

5. _____

6. _____

7. _____

8. _____

9. _____

10. _____

11. _____

12. _____

13. _____

14. _____

15. _____

16. _____

17. _____

18. _____

19. _____

20. _____

21. _____

22. _____

FIGURE 10.11   Pelvic girdle and lower limb.

E. Pelvic Girdle and Pelvis

Fill in the blank with the correct term.

_____ 1. When you put your hands on your hips, which bone marking of each os coxa are you touching?

_____ 2. With your hands on your hips, feel a point of the pelvis protruding anteriorly just above your thigh. Name this bone marking.

_____ 3. Name the two bones that form the pelvic girdle.

_____ 4. What bones of the os coxae do you sit on?

_____ 5. Name the bones of the os coxae that articulate anteriorly.

F. Lower Limb

Fill in the blank with the correct term.

_____ 1. Is the fibula medial or lateral to the tibia?

_____ 2. Medial process at the distal end of the tibia that forms the anklebone.

_____ 3. Heaviest and strongest bone of the leg (not thigh).

_____ 4. Thigh bone.

_____ 5. The fibula is a weight-bearing bone. True or False?

_____ 6. Tarsal bone that articulates with the tibia.

_____ 7. Heel bone.

_____ 8. Bones of the distal part of the instep.

JOINTS AND SYNOVIAL JOINT MOVEMENTS

OBJECTIVES

After completing this exercise, you should be able to:

- Describe the three types of structural joints and give an example of each
- Distinguish between the three types of functional joints and give an example of each
- Describe the basic structure of a typical synovial joint
- Describe the types of movements of synovial joints and demonstrate them

MATERIALS

- articulated skeleton
- whole mammalian synovial joint (fresh, frozen, or glycerinated)
- longitudinally cut mammalian synovial joint
- dissection tray and instruments
- disposable gloves
- colored pencils

A joint or **articulation** (*articulare* = to divide into joints) connects a bone with another bone, cartilage, or tooth. Joints are commonly classified according to their structure or function.

A. STRUCTURAL CLASSIFICATION OF JOINTS

The structural classification of joints depends on the type of connective tissue forming the joint, and whether or not there is a space or synovial cavity between the bones. **Fibrous joints** have dense fibrous connective tissue with strong collagen fibers that hold the joints firmly together with no synovial cavity. This type of joint has little or no movement. Examples are the skull joints, teeth in sockets, and the distal joint between the tibia and fibula. **Cartilaginous joints** have either hyaline cartilage or fibrocartilage connecting the bones with no synovial cavity. Usually, there is a small degree of movement with this type of joint. Examples are the intervertebral joints, the pubic symphysis, and the connection of the sternum and the manubrium. **Synovial joints** (*syn-* = together; *ovum* = egg) have a great amount of movement due to having a small synovial cavity (space) between the two bones. Dense fibrous connective tissue holds the bones together. The term *synovial* comes from the synovial fluid present in the synovial cavity that resembles the albumin of an uncooked egg, only more viscous. The majority of the joints in the human body are synovial joints, with examples being the shoulder, elbow, hip, and knee joints.

ACTIVITY 1 STRUCTURAL CLASSIFICATION OF JOINTS

1 Identify the structural classification of each joint in Figure 11.1.

2 Choose a color for each classification of joint. Color the leader line of each joint according to its classification.

3 Identify these joints on an articulated skeleton.

4 Pronounce the terms as you point to them on the articulated skeleton.

5 Point out these joints on your own body.

B. FUNCTIONAL CLASSIFICATION OF JOINTS

The functional classification of joints is based on the amount of movement the joint allows. Immovable joints or **synarthroses** (*syn-* = union; *arthro-* = joint) include the sutures between the skull bones and the teeth sockets. Intervertebral joints, the tibiofibular joint, and the pubic symphysis are examples of slightly movable joints or **amphiarthroses** (*amphi-* = on both sides). Most of the joints in the body, about 90%, are freely movable joints, or **diarthroses** (*di-* = apart; away from).

Most fibrous joints such as sutures and teeth sockets are nonmovable joints. However, the fibrous tibiofibular joint is a slightly movable joint. Most cartilaginous joints are slightly movable joints, such as the intervertebral discs and the pubic symphysis. The cartilaginous epiphyseal plates of long bones, however, are nonmovable joints.

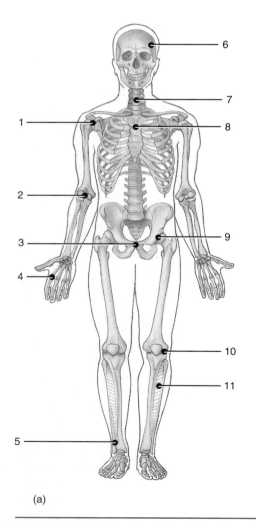

(a)

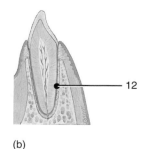

(b)

FIGURE 11.1   Structural classification of joints.

- cartilaginous joint (car-tih-LA-jih-nous)
- fibrous joint
- synovial joint (sih-NO-vi-ul)

1 _____

2 _____

3 _____

4 _____

5 _____

6 _____

7 _____

8 _____

9 _____

10 _____

11 _____

12 _____

ACTIVITY 2 FUNCTIONAL CLASSIFICATION OF JOINTS

1 Identify the functional classification of each joint in Figure 11.2.

2 Choose a color for each classification of joint. Color the leader line of each joint according to its classification.

3 Identify these joints on an articulated skeleton.

4 Pronounce the terms as you point to them on the articulated skeleton.

C. BASIC STRUCTURE OF SYNOVIAL JOINTS

Although synovial joints vary in structure, they have several common features that are listed below:

- **synovial cavity**—small space between the two articulating bones
- **articular cartilage**—hyaline cartilage covering the ends of the bones in the synovial cavity
- **articular capsule**—encloses the synovial joint and synovial cavity; has two layers, the fibrous capsule and synovial membrane
- **fibrous capsule**—outer dense fibrous connective tissue layer of the articular capsule that is continuous with the periosteum of the bone; also forms ligaments when fibrous bundles are parallel
- **synovial membrane**—inner layer of the articular capsule
- **synovial fluid**—secreted by the synovial membrane; lubricates the articular cartilages to reduce friction

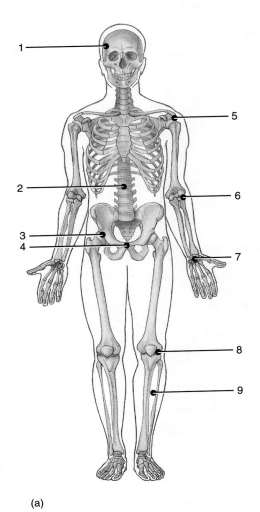

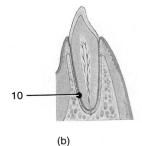

(a)   (b)

- amphiarthrosis (amphi-ar-THROW-sis)
- diarthrosis (die-ar-THROW-sis)
- synarthrosis (syn-ar-THROW-sis)

1 _____

2 _____

3 _____

4 _____

5 _____

6 _____

7 _____

8 _____

9 _____

10 _____

FIGURE 11.2   Functional classification of joints.

ACTIVITY 3 **TYPICAL SYNOVIAL JOINT**

1 Label the simple synovial joint in Figure 11.3.

2 Using contrasting colors, color the structures in Figure 11.3.

3 Pronounce each term as you complete the labeling exercise.

D. TYPES OF MOVEMENT AT SYNOVIAL JOINTS

Contraction of skeletal muscles causes bone movement at synovial joints. Professionals in kinesiology and physical therapy use particular terms to describe the movements of these joints. The four main categories of synovial joint movements are: **gliding, angular, rotation,** and **special movements** (Table 11.1). Many movements are grouped to demonstrate opposite movements.

ACTIVITY 4 **TYPES OF MOVEMENTS AT SYNOVIAL JOINTS**

1 Label each movement illustrated in Figures 11.4 through 11.7.

2 Demonstrate each movement with your partner(s).

3 Use an articulated skeleton to demonstrate each movement. During pronation, observe the radius crossing over the ulna.

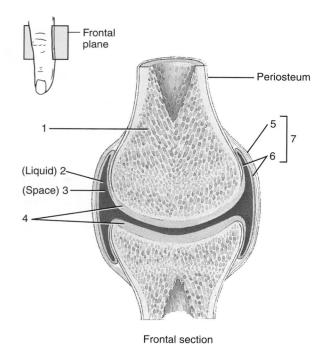

Frontal section

• articular bone (ar-TIH-ku-lar)
• articular capsule
• articular cartilage
• fibrous capsule
• synovial cavity
• synovial fluid
• synovial membrane

1 _____

2 _____

3 _____

4 _____

5 _____

6 _____

7 _____

FIGURE 11.3 **Typical synovial joint.**

TABLE 11.1 Types of Movement at Synovial Joints

MOVEMENT	DESCRIPTION
A. Gliding	Nearly flat bone surfaces slide or glide over each other
B. Angular	
Flexion (*flex-* = to bend)	Decrease in the angle of a joint
Extension (*exten-* = to stretch out)	Increase in the angle of a joint
Hyperextension (*hyper-* = excessive)	Excessive extension movement beyond normal anatomical position
Abduction (*ab-* = away; *-duct* = to lead)	Move appendage away from the midline
Adduction (*ad-* = toward)	Move appendage toward midline
Circumduction (*circ-* = circle)	Move a distal part of an appendage in a circular motion
C. Rotation (*rota-* = revolve)	Turn on a pivot with a circular motion
D. Special Joint Movements	
Elevation	Upward movement raising body part vertically
Depression	Downward movement lowering body part vertically
Protraction (*pro-* = in front of; *trahere* = to draw)	Move a body part forward or anterior on a horizontal plane
Retraction (*retractare* = to draw back)	Move a body part backward or posterior
Supination (*supine* = lying on the back)	Palm of the hand faces forward or upward
Pronation (*pronate* = lying face downward)	Palms face backward or downward
Inversion	Turn the sole of foot inward
Eversion	Turn the sole of foot outward
Dorsiflexion	Point the toes upward; stand on the heels
Plantar flexion	Point the toes downward; raise the heel

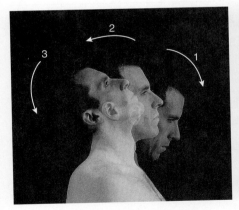

(a) Atlanto-occipital and cervical intervertebral joints

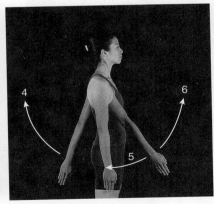

(b) Shoulder joint

(a)
• extension
• flexion
• hyperextension

1 _____

2 _____

3 _____

(b)

4 _____

5 _____

6 _____

FIGURE 11.4   **Angular joint movements of flexion, extension, and hyperextension.**

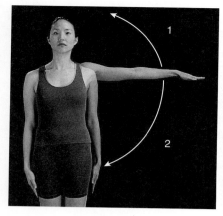

(a) Shoulder joint

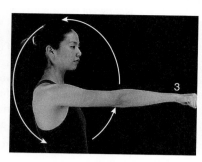

(b) Shoulder joint

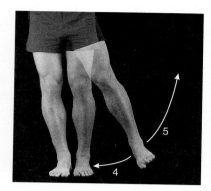

(c) Hip joint

FIGURE 11.5   **Angular joint movements of abduction, adduction, and circumduction.**

- abduction (ab-DUK-shun)
- adduction (ad-DUK-shun)
- circumduction (sir-cum-DUK-shun)

(a)

1 _____

2 _____

(b)

3 _____

(c)

4 _____

5 _____

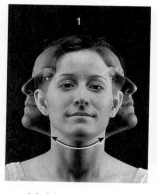

(a) Atlanto-axial joint

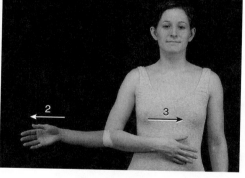

(b) Shoulder joint

- lateral rotation
- medial rotation
- rotation

1 _____

2 _____

3 _____

FIGURE 11.6   **Joint movements of rotation.**

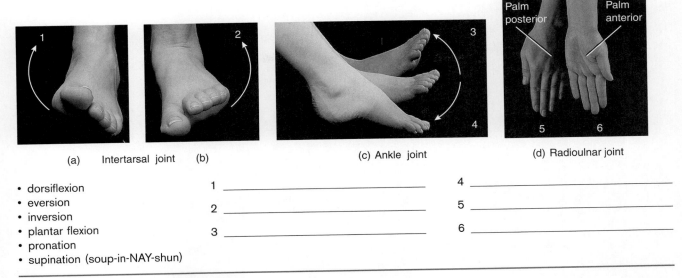

(a)   Intertarsal joint   (b)   (c) Ankle joint   (d) Radioulnar joint

- dorsiflexion
- eversion
- inversion
- plantar flexion
- pronation
- supination (soup-in-NAY-shun)

1 _____   4 _____

2 _____   5 _____

3 _____   6 _____

FIGURE 11.7   Special joint movements.

E. DISSECTION OF A SYNOVIAL JOINT

NOTE: *If you are using fresh or frozen joints, wear protective gloves and wash your hands thoroughly with soap and water after the dissection.*

ACTIVITY 5 SYNOVIAL JOINT DISSECTION

1 Dissect a whole mammalian synovial joint following the dissection instructions.

- After placing your **synovial joint** in a dissecting tray, observe the external white **articular capsule** that holds the bones together. Note that the articular capsule is continuous with the periosteum of the bone.

- Using a scalpel, cut open the joint capsule. Feel the slippery **synovial fluid** between your fingers.

- Try to detect the difference between the outer **fibrous capsule** layer and the inner **synovial membrane** layer of the articular capsule.

- Note the **synovial cavity** or space between the **articular bones.**

- Observe the ends of the joint bones for the **articular cartilage.** Cut out a small piece of hyaline cartilage to observe its thickness.

2 Pronounce the terms as you point to them on the joint.

3 Observe a demonstration of a longitudinally cut synovial joint. Repeat the dissection steps above, except do not cut the joint with a scalpel.

CLEAN UP: *Thoroughly wash your dissection tray, instruments, and work surface with soap and water. Be careful not to wash bits of tissue down the drain. Dispose of the joint and cut tissue as directed by your instructor.*

REVIEWING YOUR KNOWLEDGE

A. Structural and Functional Classification of Joints

Fill in the blank with the most appropriate term: **cartilaginous, fibrous, synovial, amphiarthrosis, diarthrosis,** or **synarthrosis.**

_____ **1.** A *functional* type of joint that has the greatest amount of movement.

_____ **2.** A *functional* type of joint that has the least amount of movement.

_____ **3.** A *functional* type of joint that is slightly movable.

_____ **4.** A *structural* type of joint that has an articular (joint) capsule.

_____ **5.** A *structural* classification that has cartilage joining the ends of the articulating bones.

_____ **6.** A *structural* type of joint with a joint cavity.

_____ **7.** A *structural* type of joint that is tightly held together by fibrous connective tissue.

_____ **8.** The *structural* classification of an intervetebral joint.

_____ **9.** The *functional* classification of an intervertebral joint.

_____ **10.** The *structural* classification of a suture.

_____ **11.** The *functional* classification of a suture.

_____ **12.** The *structural* classification of the elbow joint.

_____ **13.** The *functional* classification of the elbow joint.

B. Basic Structure of Synovial Joints

Match the description with the correct term.

a. articular cartilage

b. articular capsule

c. synovial cavity

d. synovial fluid

e. synovial membrane

_____ **1.** Structure that encloses synovial joint and synovial cavity.

_____ **2.** The fluid that lubricates, reduces friction, and gives nutrition to a joint.

_____ **3.** The tissue at the end of a bone that reduces friction in a joint.

_____ **4.** Small space between two articulating bones that reduces friction.

_____ **5.** Structure that secretes synovial fluid.

C. Movement at Synovial Joints

Fill in the blank with the correct term for the type of movement of the synovial joint.

_____ **1.** Moves appendage toward the midline.

_____ **2.** Increases the angle of a joint.

_____ **3.** Downward movement lowering the body part vertically.

_____ **4.** Palm of hand faces forward or upward.

_____ **5.** Pointing toes downward; raising the heel on your tiptoes.

_____ **6.** Decreases the angle of a joint.

_____ **7.** Moves an appendage away from the midline.

_____ **8.** Turns on a pivot with a circular motion.

_____ **9.** Movement that raises mandible.

_____ **10.** Moves body part forward along horizontal plane.

_____ **11.** Turn sole of foot inward.

CONTRACTION OF SKELETAL MUSCLE

OBJECTIVES

After completing this exercise, you should be able to:

- Describe the function of the connective tissue coverings of skeletal muscles.
- Describe the structure of the neuromuscular junction
- Describe how skeletal muscles achieve a smooth, sustained contraction
- Describe how skeletal muscles vary the force of contraction
- Define threshold of contraction, maximal stimulus, recruitment, and fatigue, and explain how to observe them

MATERIALS

- compound microscope, lens paper
- prepared microscope slides of skeletal muscle in cross- and longitudinal sections, and slides of the neuromuscular junction
- rulers and ankle weights (students may bring them from home)

Each skeletal muscle can be considered an organ because it is composed of muscle tissue, connective tissue, nerve fibers, and blood vessels. When skeletal muscles are stimulated by somatic motor neurons to contract, they will produce a variety of body motions or will stabilize body positions.

A. CONNECTIVE TISSUE COVERINGS OF SKELETAL MUSCLES

Electrical impulses that initiate a contraction in one skeletal muscle fiber are not spread to an adjacent muscle fiber.

This is due to three layers of connective tissue coverings that separate and electrically insulate skeletal muscles. Connective tissue layers also support and protect muscle fibers and extend beyond them to form tendons. **Endomysium** (*endo-* = within; *-mys* = muscle) is the inner connective tissue that covers each individual skeletal muscle fiber. Skeletal muscle fibers are grouped into bundles called **fascicles** that are surrounded by a layer of connective tissue called **perimysium** (*peri-* = around). A whole muscle is formed from a number of fascicles that are surrounded by a connective tissue layer called **epimysium** (*epi-* = on; upon).

B. THE NEUROMUSCULAR JUNCTION

Each skeletal muscle fiber is stimulated to contract by a **somatic motor neuron.** A motor neuron and all the skeletal muscle fibers it innervates are a **motor unit.** Within the muscle, the axon of a motor neuron divides into many branches or **axon terminals,** each of which forms a **neuromuscular junction** with a skeletal muscle fiber. At the neuromuscular junction, each axon terminal divides into **synaptic end bulbs.** Within the synaptic end bulbs are **synaptic vesicles** filled with neurotransmitter molecules. When a nerve impulse reaches the synaptic end bulb, neurotransmitters are released and diffuse across the **synaptic cleft** (space between the synaptic end bulb and the muscle fiber). Neurotransmitter molecules bind to receptors in the **motor end plate,** the region of the sarcolemma (plasma membrane) directly across from the synaptic end bulb. If enough neurotransmitter binds, an **action potential** is generated that stimulates the skeletal muscle fiber to contract.

ACTIVITY 1 MICROSCOPIC ANATOMY OF SKELETAL MUSCLE

1 Label the photomicrographs of sections of skeletal muscle in Figure 12.1(a) and (b).

2 Examine a microscope slide of skeletal muscle in cross-section and longitudinal section. Identify the structures listed in Figure 12.1.

3 Label the neuromuscular junction in Figure 12.2.

4 Label the photomicrograph of the neuron junction in Figure 12.3.

5 Examine a microscope slide of the neuromuscular junction and identify the structures listed in Figure 12.3.

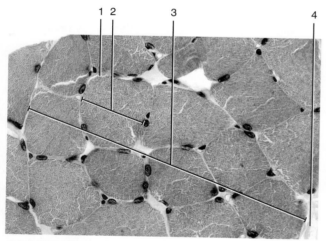

400×, H&E, human

(a) Cross-section
• endomysium
• fascicle
• perimysium
• skeletal muscle fiber in cross-section

1 _____

2 _____

3 _____

4 _____

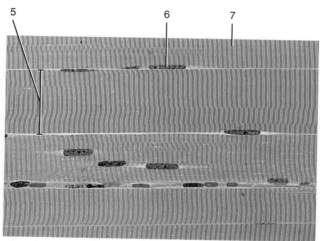

400×, H&E, human

(b) Longitudinal section
• width of skeletal muscle fiber
• nucleus
• striation (stry-AY-shun)

5 _____

6 _____

7 _____

FIGURE 12.1   Sectional views of skeletal muscle fibers.

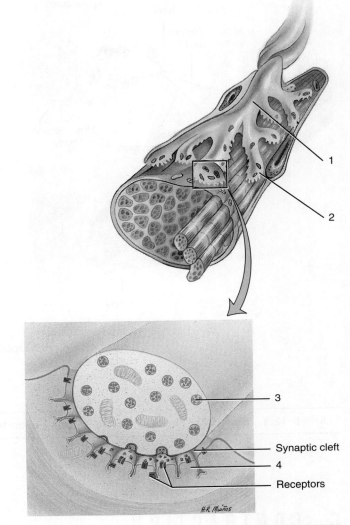

- axon terminal
- motor end plate
- synaptic end bulb
- synaptic vesicle with neurotransmitter

1 _____

2 _____

3 _____

4 _____

FIGURE 12.2 **Structure of the neuromuscular junction.**

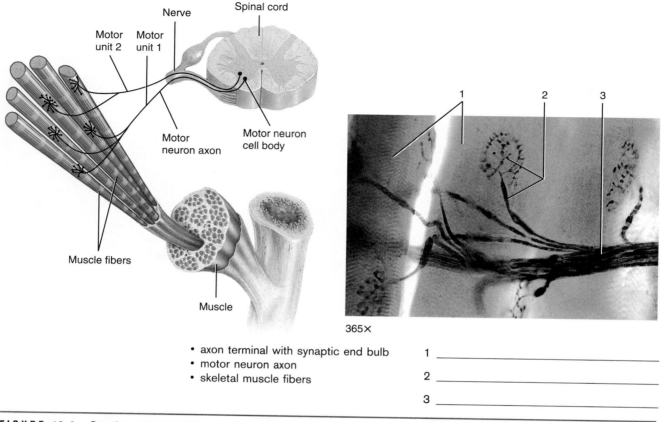

365×

- axon terminal with synaptic end bulb   1 _____
- motor neuron axon   2 _____
- skeletal muscle fibers   3 _____

FIGURE 12.3 **Section of skeletal muscle showing axon terminals and synaptic end bulbs.**

C. CONTRACTION OF SKELETAL MUSCLE FIBERS

A motor neuron stimulates all the skeletal muscle fibers in a motor unit simultaneously. The amount of force generated by a motor unit is determined by the number of muscle fibers within the motor unit, the size of the individual muscle cells, and the number of crossbridges attached in the muscle fibers.

A **twitch contraction** is the quick shortening and relaxation observed in a skeletal muscle when a single action potential traveling down a motor neuron stimulates the skeletal muscle fibers of the motor unit to contract. The three phases of a twitch contraction are the latent period, the contraction period, and the relaxation period. The **latent period** lasts about 2 milliseconds (msec) and is the time between stimulation of muscle cells and force generation. The **contraction period** lasts about 10 to 100 msec and is the period during which force (measured in grams) is increasing. The **relaxation period,** which lasts 10 to 100 msec, is the period when force is decreasing. Normal muscle contractions are not twitch contractions but are sustained contractions of varying force.

If the muscle fibers of a motor unit are stimulated before the relaxation phase of a muscle twitch is completed, then the next contraction will produce a greater force. This is called **wave summation.** Increasing the frequency of muscle stimulation produces sustained force generation. **Unfused tetanus** (*tetan* = rigid) occurs when there is a partial relaxation between muscle twitches. **Fused tetanus** is a sustained contraction with no relaxation observed between twitches.

Increasing the number of motor units contracting at the same time, **motor unit recruitment,** also increases the amount of force generated. Lifting a feather requires fewer motor units than lifting your anatomy and physiology textbook. **Maximal force** development occurs when all motor units of a muscle are stimulated and all muscle fibers are contracting.

Most sustained voluntary skeletal muscle contractions are unfused tetanic contractions with different motor units stimulated at different times (**asynchronous contractions**). The asynchronous contractions delay **muscle fatigue,** which is an inability to contract caused by long periods of muscle contraction.

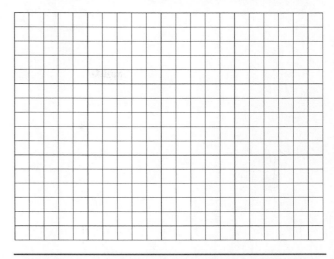

FIGURE 12.4   **Student drawing of twitch contraction.**

ACTIVITY 2 CONTRACTION OF SKELETAL MUSCLE FIBERS

1 Draw a twitch contraction that has a latent period of 2 msec, a contractile period of 10 msec with a maximum force of 2 g (grams) and a relaxation period of 10 msec. Use the graph in Figure 12.4.

2 Observe unfused tetanus, motor unit recruitment, and fatigue in the hamstring muscles.

- In your lab group, decide who will be the subject, observer(s), and recorder.
- Have the subject stand while holding onto the lab bench for support. Have the subject bend the **left** knee to a 90-degree angle and hold this position. The hamstring muscles are used to bend the knee.
- Observe the unfused tetanic contraction. Start timing.
- Time how long it takes for the muscle to fatigue, which is observed by *any* vertical movement in the bent leg.
- Time until fatigue: _____ min.
- Place a 5-lb ankle weight on the subject's **right** ankle and repeat the steps.
- Time until fatigue: _____ min.

3 Answer discussion questions with your lab partners.

❓ DISCUSSION QUESTIONS: OBSERVATION OF UNFUSED TETANUS, RECRUITMENT, AND FATIGUE IN HAMSTRING MUSCLES

1 Explain why you can maintain contraction of the hamstring muscles over time.

2 Explain why you can also sustain contraction of the hamstring muscles with a 5-lb weight on the ankle.

3 Explain why the leg with the 5-lb ankle weight fatigued faster.

REVIEWING YOUR KNOWLEDGE

A. The Neuromuscular Junction

Match the structure(s) to the definition.

    a. axon terminal
    b. motor end plate
    c. synaptic cleft
    d. synaptic end bulb
    e. synaptic vesicle

____ **1.** Found in synaptic end bulbs of axon terminal; contains neurotransmitter molecules.

____ **2.** A specialized area of sarcolemma that contains neurotransmitter receptors.

____ **3.** Space between synaptic end bulbs of axon terminal and muscle fiber.

____ **4.** Branches into synaptic end bulbs at neurotransmitter junction.

____ **5.** Part of the axon terminal that forms a neuromuscular junction.

B. Contraction of Skeletal Muscle Fibers

Match the description with the correct term.

    a. contractile period
    b. fused tetanus
    c. latent period
    d. motor unit
    e. motor unit recruitment
    f. muscle fatigue
    g. relaxation period
    h. twitch
    i. unfused tetanus

_____ **1.** Inability of muscle fibers to contract after a long period of contraction.

_____ **2.** Phase of contraction; time between muscle fiber stimulation and measurement of force generation.

_____ **3.** Motor neuron and all the muscle fibers it innervates.

_____ **4.** Phase of contraction; period of force generation.

_____ **5.** Type of wave summation with partial relaxation observed between twitches.

_____ **6.** Increasing the number of motor units that are stimulated to contract.

_____ **7.** Single contractile event in response to single action potential.

_____ **8.** Phase of contraction; period during which more crossbridges detach than reattach to thin filaments.

_____ **9.** Type of wave summation with no observable relaxation between twitches.

SKELETAL MUSCLES AND THEIR ACTIONS

OBJECTIVES

After completing this exercise, you should be able to:

- Describe how skeletal muscles are named
- Identify major skeletal muscles on models or charts
- Describe the actions of the major skeletal muscles

MATERIALS

- muscle models, anatomical charts, and/or cadaver photographs
- colored pencils (optional)

A. CRITERIA FOR NAMING SKELETAL MUSCLES

Learning the criteria used to name skeletal muscles will assist you in identifying and learning the function of these muscles. Skeletal muscles are named for their orientation relative to the midline of the body, size, shape, action, number of origins, location, or origin and insertion. The muscle attachment points, the origin and insertion, are usually skin or bone. The **origin** is the nonmoving point of attachment when a muscle contracts and the **insertion** point moves toward the origin. Table 13.1 identifies criteria for naming skeletal muscles.

ACTIVITY 1 CRITERIA FOR NAMING SKELETAL MUSCLES

Each word in a skeletal muscle name gives information about the muscle. Give the criteria used for each word below.

Name	Criteria
1 orbicularis oculi	orbicularis—direction of fibers (circular) oculi—location (eye)
2 biceps brachii	_____
3 rectus femoris	_____
4 gluteus maximus	_____
5 deltoid	_____
6 sternocleidomastoid	_____

TABLE 13.1 Criteria for Naming Skeletal Muscles

NAME	DEFINITION	NAME	DEFINITION
ORIENTATION OF MUSCLE RELATIVE TO BODY MIDLINE		**MUSCLE ACTION**	
Rectus	Muscle parallel to body midline	Flexor	Decreases joint angle
Transverse	Muscle perpendicular to body midline	Extensor	Increases joint angle
Oblique	Muscle diagonal to body midline	Abductor	Moves bone away from midline
		Adductor	Moves bone close to midline
MUSCLE SIZE		Levator	Elevates
Maximus	Largest	Depressor	Depresses
Medius	Middle	Supinator	Turns palm superiorly or anteriorly
Minimus	Smallest	Pronator	Turns palm inferiorly or posteriorly
Magnus	Large	Sphincter	Decreases size of opening
Brevis	Shortest	Tensor	Makes a body part rigid
Latissimus	Wide	Rotator	Moves bone around longitudinal axis
Major	Larger		
Minor	Smaller	**NUMBER OF ORIGINS**	
Vastus	Great	Biceps	Two origins
		Triceps	Three origins
MUSCLE SHAPE		Quadriceps	Four origins
Deltoid	Triangular		
Trapezius	Trapezoid	**LOCATION**	
Serratus	Saw-toothed	Frontalis	Located on the frontal bone
Rhomboid	Diamond-shaped		
Orbicularis	Circular		
Pectinate	Comb-like	**ORIGIN AND INSERTION**	
Piriformis	Pear-shaped	Sternohyoid	Originates on the sternum; inserts on the hyoid bone
Platysma	Flat		
Quadratus	Square		
Gracilis	Slender		

B. SKELETAL MUSCLE IDENTIFICATION AND ACTION

Word derivatives and the muscle location are provided to help you identify the muscle. Within a location, some muscles are grouped according to whether they are **superficial** or **deep.**

When skeletal muscles contract, they pull their point of insertion toward their point of origin. For the muscles of facial expression, this moves the skin on the face and allows us to express emotion. For most other skeletal muscles, contraction results in movement of a bone at a joint. In the appendages, muscles that perform similar movements are surrounded by fascia, forming **compartments** that are served by the same major blood vessels and nerves.

Some skeletal muscles have many actions. This exercise, however, only includes the *main action* of a muscle that differentiates it or characterizes it as part of a contracting group. Some muscles cross two joints, and actions at both joints are included for these muscles.

1. Muscles of the Head and Neck

Muscles of Facial Expression and Mastication

The muscles of facial expression move the skin of the face to convey a variety of emotions. Muscles of mastication move the mandible and help hold food between the teeth while chewing.

Muscles that Move the Head and Neck

Muscles that flex the head and neck are found on the anterior surface of the neck. These muscles exist in pairs, one on each side of the neck. Contraction of both muscles causes flexion while contraction of one turns the head to the side. Muscles that extend the head and neck are found on the posterior surface of the neck.

ACTIVITY 2 SKELETAL MUSCLES OF THE HEAD AND NECK

1 Label Figure 13.1(a) and (b).

2 Use Figure 13.1 and Table 13.2 to point to the muscles on models, charts, and/or cadaver photographs. Pronounce the names as you point to them.

3 Locate the muscles on yourself and perform the actions indicated in Table 13.2. Palpate the muscles while performing the actions.

4 *Optional:* Color the muscles in Figure 13.1 in contrasting colors. Color in the same direction as the muscle fibers.

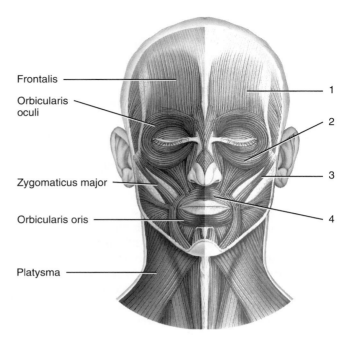

Frontalis

Orbicularis oculi

Zygomaticus major

Orbicularis oris

Platysma

1
2
3
4

- frontalis (fron-TA-lis)
- orbicularis oculi (or-bi-kyoo-LAR-is OK-yoo-lie)
- orbicularis oris (OR-is)
- zygomaticus major (zy-go-MA-ti-kus)

1 _____
2 _____
3 _____
4 _____

(a) Anterior superficial view

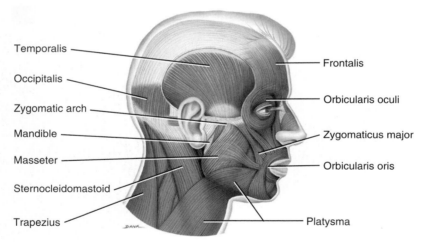

Temporalis

Occipitalis

Zygomatic arch

Mandible

Masseter

Sternocleidomastoid

Trapezius

Frontalis

Orbicularis oculi

Zygomaticus major

Orbicularis oris

Platysma

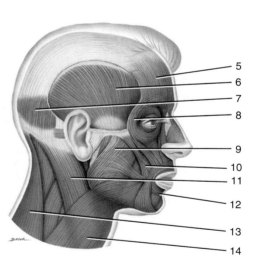

5
6
7
8
9
10
11
12
13
14

(b) Right lateral superficial view

- frontalis
- masseter (MAS-se-ter)
- occipitalis (ok-si-pi-TA-lis)
- orbicularis oculi
- orbicularis oris
- platysma
- sternocleidomastoid
- temporalis (tem-por-A-lis)
- trapezius
- zygomaticus major

5 _____
6 _____
7 _____
8 _____
9 _____

10 _____
11 _____
12 _____
13 _____
14 _____

FIGURE 13.1 Selected muscles of the head.

TABLE 13.2 Muscles of the Head and Neck

MUSCLE	LOCATION AND FUNCTION
MUSCLES OF FACIAL EXPRESSION	(FUNCTION IN ITALICS)
Frontalis (*frontalis* = in front)	Lies over frontal bone. *Raises eyebrows and wrinkles forehead.*
Occipitalis (*occipito* = base of skull)	Lies over occipital bone. *Pulls scalp posteriorly.*
Orbicularis oculi (*orbicularis* = little circle; *oculi* = eye)	Encircles the eye; circular muscle. *Closes eye.*
Zygomaticus major (*zygoma* = bar)	Between zygomatic bone and corner of mouth. *Raises corners of mouth (smiling).*
Orbicularis oris (*oris* = mouth)	Encircles the mouth; circular muscle. *Closes and purses lips.*
MUSCLES OF MASTICATION	
Temporalis (*tempora* = temples)	Lies over temporal bone. *Elevates and retracts mandible.*
Masseter (*maseter* = chewer)	Between zygomatic arch and posterior portion of mandible. *Elevates and retracts mandible.*
MUSCLES THAT MOVE THE HEAD	
Sternocleidomastoid or **SCM** (*sterno-* = breastbone; *cleido* = clavicle; *mastoid* = mastoid process)	On anterior and lateral neck; fibers run diagonally across the neck between sternum, clavicle, and mastoid process; strap-like muscle. *Both muscles contracting flex head (prayer muscle). One muscle contracting rotates head side to side (as in saying no).*
Trapezius (superior portion)	Superficial muscle; upper back and posterior neck. *Extends head.*

2. Muscles of the Trunk

Muscles that Move the Arm at the Shoulder Joint

These muscles originate on the trunk and insert on the humerus. Contraction pulls the humerus toward the insertion, resulting in movement of the humerus at the shoulder joint. Because of the structure of the shoulder joint and the way these muscles surround the humerus, contraction can cause the full range of motion possible at a synovial joint—flexion, extension, abduction, adduction, circumduction, and rotation. Muscles that insert on the anterior surface of the arm flex the arm while those on the posterior surface extend the arm. Muscles that insert on the lateral surface of the humerus abduct the arm while those on the medial surface adduct the arm.

Muscles that Move the Scapula

Movements of the scapula include elevation, depression, abduction, adduction, and rotation. Shrugging the shoulders or lifting an object over the head elevates the scapula. Depression of the scapula occurs when doing a pull-up. Doing a push-up abducts the scapula. Standing at attention adducts the scapula.

Muscles that Move the Abdominal Wall and Vertebral Column

Most abdominal muscles flex the vertebral column, while the muscles of the back extend it. The transversus abdominis muscle cannot flex the vertebral column because it runs horizontally.

ACTIVITY 3 SKELETAL MUSCLES OF THE TRUNK

1 Label Figure 13.2(a) and (b) and Figure 13.3.

2 Use Figures 13.2 and 13.3 and Tables 13.3 and 13.4 to point to the muscles on models, charts, and/or cadaver photographs. Pronounce the names as you point to them.

3 Locate each muscle on yourself and perform the action indicated in Tables 13.3 and 13.4. Palpate the muscle while performing the action.

4 *Optional:* Color the muscles in Figures 13.2 and 13.3 in contrasting colors. Color in the same direction as the muscle fibers.

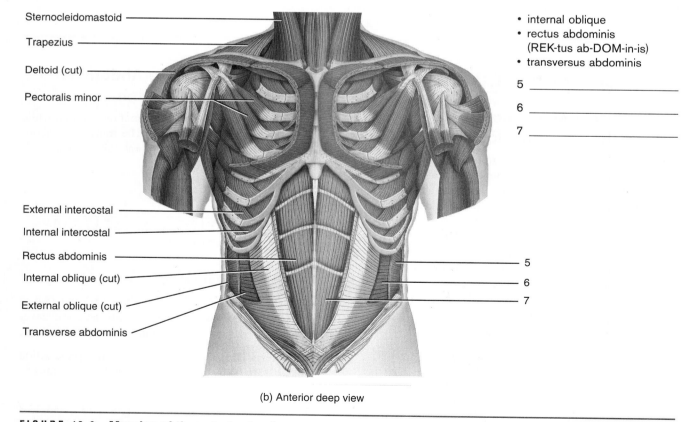

Platysma

Clavicle

Deltoid

Pectoralis major

Latissimus dorsi

Serratus anterior

Linea alba

External oblique

1

2

3

4

- deltoid
- external oblique
- pectoralis major (pek-tor-A-lis)
- serratus anterior (ser-RAY-tus)

1 _____

2 _____

3 _____

4 _____

(a) Anterior superficial view

Sternocleidomastoid

Trapezius

Deltoid (cut)

Pectoralis minor

External intercostal

Internal intercostal

Rectus abdominis

Internal oblique (cut)

External oblique (cut)

Transverse abdominis

5

6

7

- internal oblique
- rectus abdominis (REK-tus ab-DOM-in-is)
- transversus abdominis

5 _____

6 _____

7 _____

(b) Anterior deep view

FIGURE 13.2   Muscles of the anterior trunk.

TABLE 13.3 Muscles of the Anterior Trunk

MUSCLE	LOCATION AND FUNCTION
MUSCLES THAT MOVE THE ARM AT THE SHOULDER JOINT	(FUNCTION IN ITALICS)
Deltoid (*delta* = triangle)	Large, rounded, triangular shoulder muscle. *Anterior portion flexes arm; lateral portion abducts arm; posterior portion extends arm.*
Pectoralis major (*pectus* = chest)	Large muscles on superior, anterior chest between sternum and arm. *Adducts and flexes arm at shoulder joint.*
MUSCLES THAT MOVE THE SCAPULA	
Serratus anterior (*serratus* = saw-toothed)	Serrated-looking muscles on the lateral trunk inferior to arms and ribs. *Abducts scapula and rotates it upward. This occurs when throwing a punch and is often called the boxer's muscle.*
MUSCLES THAT MOVE THE ABDOMINAL WALL	
Superficial Muscles	
External oblique	Lateral and anterior sheet-like abdominal muscles whose fibers run obliquely toward the midline. *Both muscles contracting flex vertebral column and compress abdomen. One muscle contracting laterally flexes vertebral column.*
Rectus abdominis (*rectus* = parallel fibers)	Midline abdominal muscles located between sternum and the groin. *Flexes vertebral column and compresses abdomen. Does not laterally flex vertebral column.*
Deep Muscles	
Internal oblique	Abdominal muscles deep to external oblique whose fibers run obliquely toward the midline. *Both muscles contracting flex vertebral column and compress abdomen. One muscle contracting laterally flexes vertebral column.*
Transversus abdominis	Abdominal muscles deep to internal oblique whose fibers run transversely. *Compresses abdomen only.*

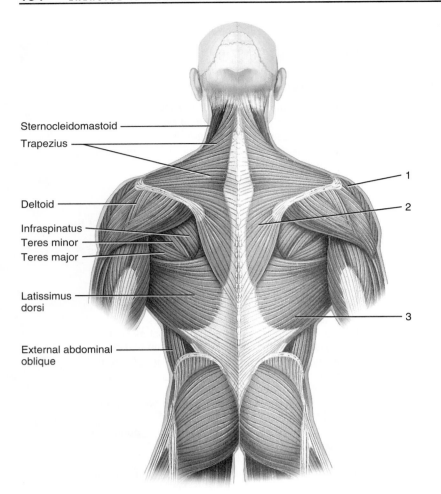

Sternocleidomastoid

Trapezius

Deltoid

Infraspinatus

Teres minor

Teres major

Latissimus dorsi

External abdominal oblique

1

2

3

Posterior superficial view

- deltoid
- latissimus dorsi (la-TIS-i-mus DOR-sigh)
- trapezius (tra-PEE-zee-us)

1 _____

2 _____

3 _____

FIGURE 13.3 **Muscles of the posterior trunk.**

TABLE 13.4 Muscles of the Posterior Trunk

MUSCLE	LOCATION AND FUNCTION
MUSCLES THAT MOVE THE ARM AT THE SHOULDER JOINT	(FUNCTION IN ITALICS)
Deltoid (*delta* = triangle)	Large, rounded, triangular shoulder muscle. *Posterior portion of deltoid extends arm.*
Latissimus dorsi (*latissimus* = widest; *dorsum* = back)	Main large, flat muscle of middle and lower back that runs laterally and attaches to superior portion of humerus. *Extends and adducts arm at shoulder joint (if arm is elevated over head, it brings it down).*
MUSCLE THAT MOVES THE SCAPULA	
Trapezius (*trapezoides* = trapezoid shape)	Diamond-shaped muscle of posterior neck and upper back that extends from the skull to spine of scapula to vertebral column. *Superior portion elevates scapula, middle portion adducts scapula, inferior portion depresses scapula.*
MUSCLE THAT MOVES THE VERTEBRAL COLUMN	
Erector spinae (*erector* = erect posture; *spinae* = spinal column)	Group of deep muscles next to the vertebral column and deep to the trapezius, latissimus dorsi, and scapula (not shown on Figure 13.3). *Extends vertebral column and maintains erect posture when both muscles contract. Laterally flexes vertebral column when only one muscle contracts.*

3. Muscles of the Arm and Forearm

Muscles located in the arm move the forearm at the elbow joint and at the radioulnar joints. These muscles have their origin on the scapula or humerus, cross the elbow joint, and insert on the bones of the forearm. When these muscles contract, they flex or extend the forearm or cause movement at the radioulnar joints. Muscles that have origins on the scapula can also move the arm at the shoulder joint.

Muscles of the appendages that perform similar functions often are surrounded by fascia, which forms compartments. Muscles in the **anterior compartments** of the arm, such as the biceps brachii and brachialis muscles, flex the forearm whereas those in the **posterior compartment,** the triceps brachii, extend the forearm.

4. Muscles of the Forearm

Most muscles of the forearm originate on the humerus, ulna, or radius and cross the wrist to insert on the hand or digits. These muscles flex and extend the wrist and digits.

The flexors are found on the anterior surface of the forearm, whereas the extensors are found on the posterior surface of the forearm. The brachioradialis, a large muscle found in the forearm, originates on the humerus and crosses the elbow joint to insert on the radius to flex the forearm.

ACTIVITY 4 SKELETAL MUSCLES OF THE ARM AND FOREARM

1 Label Figure 13.4(a) and (b) and Figure 13.5(a) and (b).

2 Use Figures 13.4 and 13.5 to point to the muscles on models, charts, and/or cadaver photographs. Pronounce the names as you point to them.

3 Locate each muscle on yourself and perform the action indicated in Tables 13.5 and 13.6. Palpate the muscle while performing the action.

4 *Optional:* Color the muscles in Figures 13.4 and 13.5 in contrasting colors. Color in the same direction as the muscle fibers.

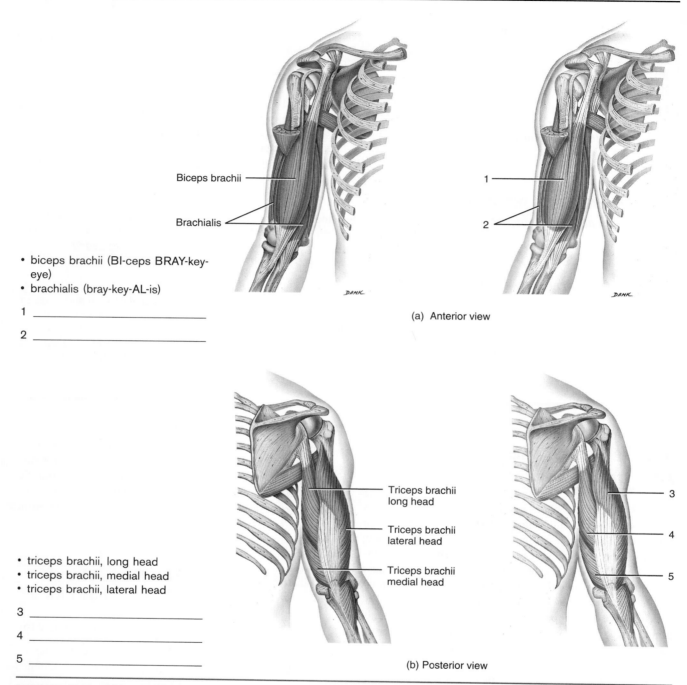

Biceps brachii

Brachialis

- biceps brachii (BI-ceps BRAY-key-eye)
- brachialis (bray-key-AL-is)

1 _____

2 _____

(a) Anterior view

Triceps brachii long head

Triceps brachii lateral head

Triceps brachii medial head

- triceps brachii, long head
- triceps brachii, medial head
- triceps brachii, lateral head

3 _____

4 _____

5 _____

(b) Posterior view

FIGURE 13.4   Muscles of the arm.

- brachioradialis (BRAY-kee-oh-ray-dee-AL-is)
- flexor carpi radialis (KAR-pee)
- flexor carpi ulnaris
- flexors

1 _____

2 _____

3 _____

4 _____

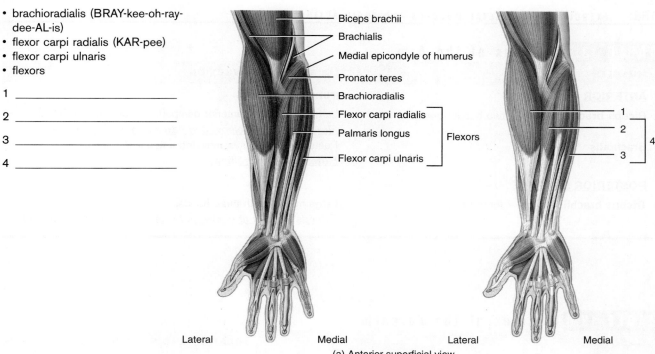

(a) Anterior superficial view

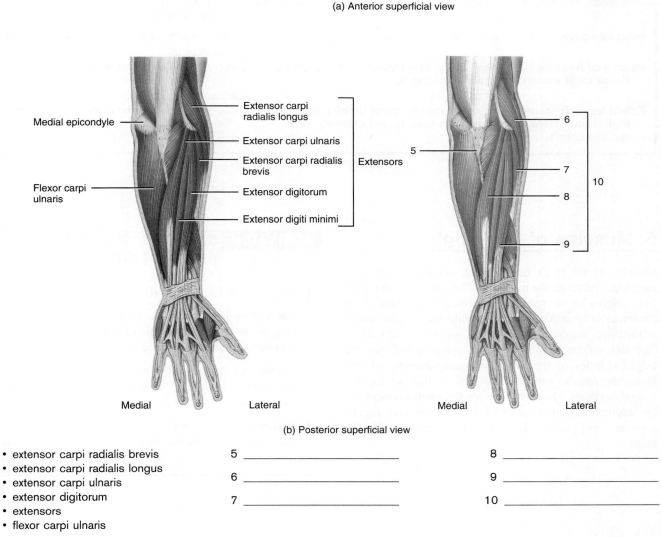

(b) Posterior superficial view

- extensor carpi radialis brevis
- extensor carpi radialis longus
- extensor carpi ulnaris
- extensor digitorum
- extensors
- flexor carpi ulnaris

5 _____

6 _____

7 _____

8 _____

9 _____

10 _____

FIGURE 13.5 Muscles of the forearm, wrist, and hands.

TABLE 13.5 Muscles of the Arm

MUSCLE	LOCATION AND FUNCTION
ANTERIOR SURFACE	**(FUNCTION IN ITALICS)**
Biceps brachii (*biceps* = two heads; *brachii* = arm)	Largest muscle in anterior compartment of arm; two heads. *Flexes forearm at elbow and flexes arm at shoulder joint.*
Brachialis	Posterior to biceps brachii and anterior to humerus. *Flexes forearm at elbow.*
POSTERIOR SURFACE	
Triceps brachii (*triceps* = three heads)	Large muscle with three heads. *Extends forearm at elbow and extends arm at shoulder joint.*

TABLE 13.6 Muscles of the Forearm

MUSCLE	LOCATION AND FUNCTION
	(FUNCTION IN ITALICS)
Brachioradialis (*brachi-* = arm; *radialis* = radius)	Anterior forearm; large, superficial lateral muscle. *Flexes forearm.*
Flexors of forearm Includes the **flexor carpi** (*carpi* = wrist) **radialis, flexor carpi ulnaris,** and **palmaris longus.**	Group of muscles on the anterior surface of the forearm. *Flex hand at wrist joint.*
Extensors of forearm Includes the **extensor carpi ulnaris, extensor digitorum, extensor carpi radialis longus,** and **extensor carpi radialis brevis.**	Group of muscles on the posterior surface of the forearm. *Extend hand at wrist joint.*

5. Muscles of the Thigh

Muscles of the thigh flex, extend, abduct, adduct, circumduct, and rotate the thigh. Muscles that move the thigh have origins on the pelvis or vertebrae and insertions on the femur or tibia. Muscles of the **anterior compartment** of the thigh, the quadriceps femoris group, extend the leg. The rectus femoris muscle of this group also flexes the thigh. Muscles of the **posterior compartment** are the hamstring muscle group. These muscles flex the leg and extend the thigh. Muscles of the **medial compartment** are the adductor magnus, adductor longus, adductor brevis, pectineus, and gracilis. These muscles adduct and flex the thigh.

ACTIVITY 5 SKELETAL MUSCLES OF THE THIGH

1 Label Figure 13.6(a) and (b).

2 Use Figure 13.6 and Table 13.7 to point to the muscles on models, charts, and/or cadaver photographs. Pronounce the names as you point to them.

3 Locate each muscle on yourself and perform the action indicated in Table 13.7. Palpate the muscle while performing the action.

4 *Optional:* Color the muscles in Figure 13.6 in contrasting colors. Color in the same direction as the muscle fibers.

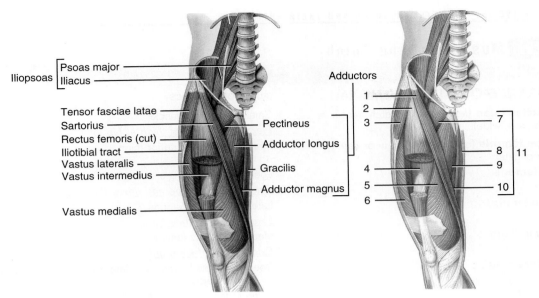

(a) Anterior superficial view

- adductors
- adductor longus
- adductor magnus
- gracilis (grah-SIL-us)
- pectineus (pek-TIN-ee-us)
- rectus femoris (FEM-or-is)
- sartorius (sar-TOR-ee-us)
- tensor fasciae latae (FA-schee LA-tee)
- vastus intermedius
- vastus lateralis
- vastus medialis

1 _____   7 _____

2 _____   8 _____

3 _____   9 _____

4 _____   10 _____

5 _____   11 _____

6 _____

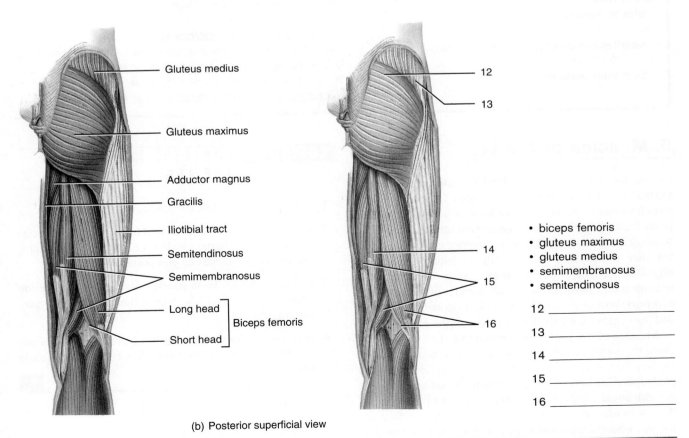

- biceps femoris
- gluteus maximus
- gluteus medius
- semimembranosus
- semitendinosus

12 _____

13 _____

14 _____

15 _____

16 _____

(b) Posterior superficial view

FIGURE 13.6   Muscles of the thigh.

TABLE 13.7 Muscles of the Thigh

MUSCLE	LOCATION AND FUNCTION
ANTERIOR SURFACE (LATERAL TO MEDIAL)	(FUNCTION IN ITALICS)
Tensor fasciae latae (*tensor* = to stretch; *fascia* = band; *latus* = wide)	Small lateral hip muscle. *Flexes and abducts thigh at hip joint.*
Quadriceps femoris (*quad* = four; *cep-* or *caput* = head or origins)	Four muscles of anterior thigh.
Rectus femoris	Located along midline of thigh. *Extends leg at knee and flexes thigh at hip.*
Vastus lateralis	Lateral to rectus femoris. *Extends leg at knee only.*
Vastus medialis	Medial to rectus femoris. *Extends leg at knee only.*
Vastus intermedius	Deep to rectus femoris and intermediate to vastus lateralis and vastus medialis. *Extends leg at knee only.*
Sartorius (*sartor* = tailor)	Diagonal muscle running from tensor fascia latae to beyond medial knee. *Flexes leg at knee and flexes, abducts, and laterally rotates thigh at hip (allows us to flex and cross our legs).*
Adductors—Group of muscles on the medial surface of the thigh. Includes the pectineus, adductor magnus, adductor longus, adductor brevis, and gracilis.	*Adduct thigh at hip joint.*
POSTERIOR SURFACE	
Gluteus maximus (*gluteus* = buttocks)	Largest buttocks muscle. *Extends thigh at hip joint only.*
Hamstrings	Three muscles of posterior thigh.
Biceps femoris	Most lateral hamstring. *Flexes leg at knee and extends thigh at hip.*
Semitendinosus (*semi-* = one half; *tendinosus* = long tendon)	Medial to biceps femoris and superficial to semimembranosus. *Flexes leg at knee and extends thigh at hip.*
Semimembranosus	Medial to biceps femoris and deep to semitendinosus. *Flexes leg at knee and extends thigh at hi.p*

6. Muscles of the Leg

Palpate the tibia of your leg. Find the anterior crest and the lateral and medial borders of the tibia. Observe that the muscles anterior to the tibia are located lateral to the anterior crest of the tibia. The anterior and lateral superficial muscles are in order starting at the anterior crest just below the knee and moving laterally: tibialis anterior, extensor digitorum longus, and peroneus longus. The tibialis anterior and extensor digitorum are muscles of the **anterior compartment** of the leg. These muscles dorsiflex the foot and also extend the toes if they insert on the phalanges.

Muscles of the **posterior compartment** of the leg include the gastrocnemius and soleus. These muscles plantar flex the foot and also flex the toes if they insert on the phalanges. Tendons of the gastrocnemius and soleus form the **calcaneal (Achilles) tendon.** Eversion of the foot is due to contraction of muscles near the lateral surface of the leg, whereas inversion is due to contraction of muscles near the medial surface of the leg.

ACTIVITY 6 SKELETAL MUSCLES OF THE LEG

1 Label Figure 13.7(a), (b), and (c).

2 Use Figure 13.7 and Table 13.8 to point to the muscles on models, charts, and/or cadaver photographs. Pronounce the names as you point to them.

3 Locate each muscle on yourself and perform the action indicated in Table 13.8. Palpate the muscle while performing the action.

4 *Optional:* Color the muscles in Figure 13.7 in contrasting colors. Color in the same direction as the muscle fibers.

- extensor digitorum longus
- flexor digitorum longus
- gastrocnemius
  (gas-trok-NEE-mee-us)
- fibularis (peroneus) longus
- soleus (SOLE-ee-us)
- tibialis anterior

1 _____

2 _____

3 _____

4 _____

5 _____

6 _____

Tibialis anterior

Extensor digitorum longus

Fibularis (peroneus) longus

Gastrocnemius

Soleus

Flexor digitorum longus

1 — 4

2 — 5

3 — 6

(a) Anterior superficial view

- extensor digitorum longus
- gastrocnemius
- fibularis (peroneus) longus
- soleus
- tibialis anterior

7 _____

8 _____

9 _____

10 _____

11 _____

Gastrocnemius

Soleus

Fibularis (peroneus) longus

Extensor digitorum longus

Tibialis anterior

7
8
9
10
11

(b) Lateral superficial view

FIGURE 13.7 **Muscles of the leg, ankle, and foot.**

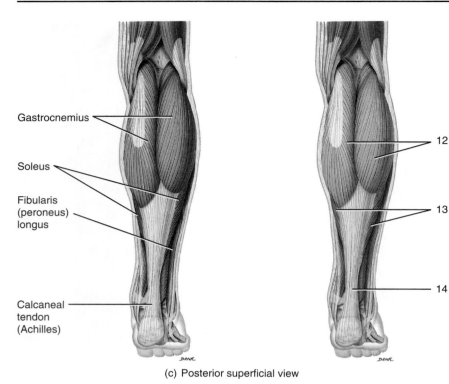

- calcaneal tendon
- gastrocnemius
- soleus

12 _____

13 _____

14 _____

(c) Posterior superficial view

FIGURE 13.7 Muscles of the leg, ankle, and foot, *continued*.

TABLE 13.8 Muscles of the Leg

MUSCLE	LOCATION AND FUNCTION
ANTERIOR AND LATERAL SUPERFICIAL MUSCLES	*(FUNCTION IN ITALICS)*
Tibialis anterior	Lateral to the anterior crest of the tibia. *Dorsiflexes and inverts foot.*
Extensor digitorum longus	Lateral to the tibialis anterior. *Dorsiflexes foot and extends toes.*
Fibularis (peroneus) longus	Lateral to the extensor digitorum longus. *Plantar flexes and everts foot.*
POSTERIOR SURFACE MUSCLES	
Gastrocnemius	Large superficial muscle on posterior leg. *Plantar flexes foot and flexes leg at knee.*
Soleus	Deep to gastrocnemius and posterior to the fibularis (peroneus) longus. *Plantar flexes foot only.*

REVIEWING YOUR KNOWLEDGE

A. Criteria for Naming Muscles

Identify the criteria used to name each muscle. More than one criteria may apply to a muscle name.

a. orientation of muscle relative to body midline
b. muscle size
c. muscle shape
d. muscle action
e. number of origins
f. location
g. origin and insertion

_____ **1.** vastus lateralis

_____ **2.** trapezius

_____ **3.** adductors

_____ **4.** sternocleidomastoid

_____ **5.** biceps brachii

_____ **6.** gluteus maximus

_____ **7.** rectus abdominis

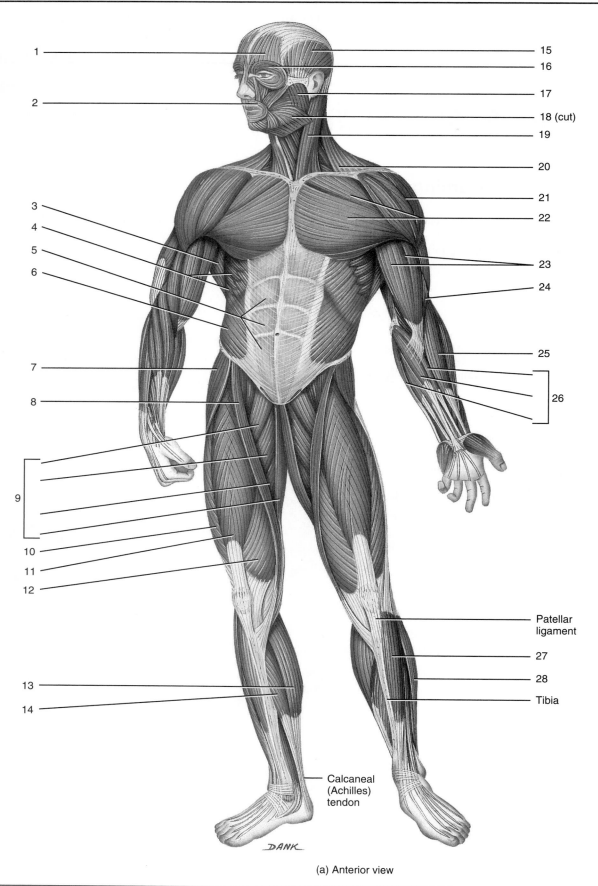

1
2
3
4
5
6
7
8
9
10
11
12
13
14

15
16
17
18 (cut)
19
20
21
22
23
24
25
26
27
28

Patellar
ligament

Tibia

Calcaneal
(Achilles)
tendon

DANK

(a) Anterior view

FIGURE 13.8 Superficial skeletal muscles.

B. Muscle Location: Superficial Skeletal Muscles

Label the muscles in Figures 13.8.

(a) Anterior view

1. _____

2. _____

3. _____

4. _____

5. _____

6. _____

7. _____

8. _____

9. _____

10. _____

11. _____

12. _____

13. _____

14. _____

15. _____

16. _____

17. _____

18. _____

19. _____

20. _____

21. _____

22. _____

23. _____

24. _____

25. _____

26. _____

27. _____

28. _____

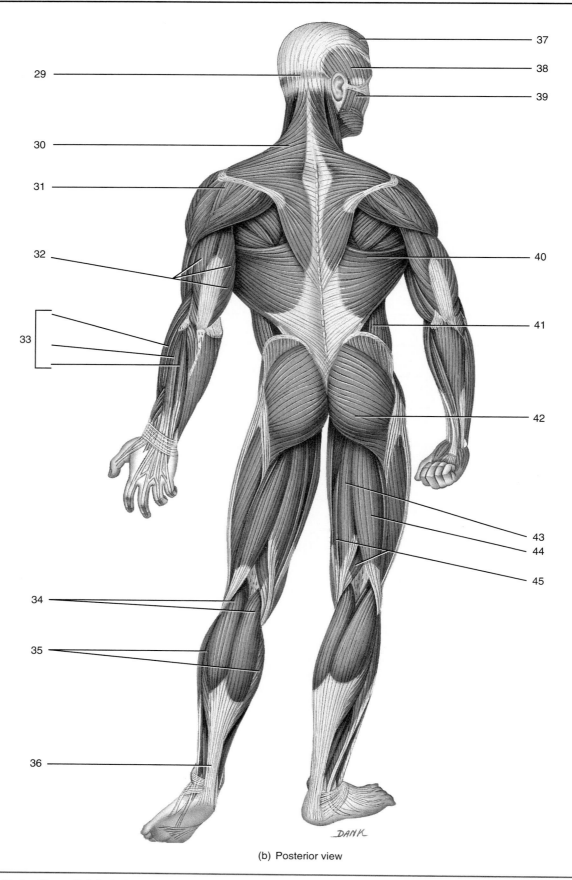

(b) Posterior view

FIGURE 13.8 **Superficial skeletal muscles,** *continued.*

(b) Posterior view

29. _____

30. _____

31. _____

32. _____

33. _____

34. _____

35. _____

36. _____

37. _____

38. _____

39. _____

40. _____

41. _____

42. _____

43. _____

44. _____

45. _____

C. Muscle Function

Identify the muscles that perform each function.

Muscles of the Head and Neck
masseter
orbicularis oculi
orbicularis oris
sternocleidomastoid
temporalis
trapezius
zygomaticus major

_____ **1.** Smiling muscle

_____ **2.** Kissing muscle

_____ **3.** Closes eyelid

_____ **4.**
                         } Closes mouth (elevates mandible)
_____ **5.**

_____ **6.** Both muscles contracting flex head; one muscle contracting rotates head side to side

_____ **7.** Extends head

Muscles of the Anterior Trunk

deltoid
external oblique
internal oblique
pectoralis major
rectus abdominis
serratus anterior
transversus abdominis

_____ **8.** Adducts and flexes arm at shoulder joint.

_____ **9.** Anterior portion flexes arm, lateral portion abducts arm.

_____ **10.** Abducts scapular and rotates it upward (boxer's muscle).

_____ **11.** Flexes vertebral column and compresses abdomen.

_____ **12.** ⎫
⎬ Both muscles contracting flex vertebral column and compress abdomen; one
⎪ muscle contracting laterally flexes vertebral column.
_____ **13.** ⎭

_____ **14.** Compresses abdomen only.

Muscles of the Posterior Trunk

deltoid
erector spinae
latissimus dorsi
trapezius

_____ **15.** Extends and adducts arm at shoulder joint.

_____ **16.** Posterior portion extends arm and lateral portion abducts arm.

_____ **17.** Superior portion extends head and elevates scapula, middle portion adducts scapula, inferior portion depresses scapula.

_____ **18.** Extends vertebral column and maintains erect posture when both muscles contract; one muscle contracting laterally flexes vertebral column.

Muscles of the Arm and Forearm

biceps brachii
brachialis
brachioradialis
extensors
flexors
triceps brachii

_____ **19.** Extends forearm at elbow and extends arm at shoulder.

_____ **20.** Flexes forearm at elbow and flexes arm at shoulder.

_____ **21.**
                    } Flexes forearm at elbow.

_____ **22.**

_____ **23.** Flex hand at wrist.

_____ **24.** Extend hand at wrist.

Muscles of the Thigh

adductors
biceps femoris
gluteus maximus
rectus femoris
sartorius
semimembranosus

semitendinosus
tensor fasciae latae
vastus intermedius
vastus lateralis
vastus medialis

_____ **25.** Extends leg at knee and flexes thigh at hip.

_____ **26.**

_____ **27.** } Extends leg at knee only.

_____ **28.**

_____ **29.** Flexes leg at knee and flexes, abducts, and laterally rotates thigh at hip (allows us to flex and cross our legs).

_____ **30.** Group of muscles that adduct and flex thigh at hip.

_____ **31.** Abducts thigh at hip.

_____ **32.** Extends thigh at hip.

_____ **33.**

_____ **34.** } Flexes leg at knee and extends thigh at hip.

_____ **35.**

Muscles of the Leg
extensor digitorum longus
fibularis (peroneus) longus
gastrocnemius
soleus
tibialis anterior

_____   **36.** Plantar flexes and everts foot.

_____   **37.** Plantar flexes foot only.

_____   **38.** Plantar flexes foot and flexes leg at knee.

_____   **39.** Dorsiflexes foot and extends toes.

_____   **40.** Dorsiflexes and inverts foot.

14

SPINAL CORD AND SPINAL NERVES

OBJECTIVES

After completing this exercise, you should be able to:

- Identify the external and internal features of the spinal cord on models or charts
- Name the four spinal divisions of a spinal cord
- Name the spinal plexuses and the major nerves arising from each plexus, and identify them on a model or chart

MATERIALS

- colored pencils
- model or chart of the spinal cord, transverse section of the spinal cord, and vertebral column with spinal cord and spinal nerves
- model or chart with major nerves of the upper and lower extremities
- prepared microscope slides or photomicrographs of spinal cord transverse section
- preserved or fresh spinal cord with meninges (for demonstration)
- dissection equipment, dissection tray, disposable gloves

The spinal cord and the brain comprise the central nervous system. Being continuous with the brain, the spinal cord begins at the foramen magnum and terminates between vertebrae L1 and L2. It is suspended within the vertebral canal, an area formed by the vertebral foramen of the vertebral column. The spinal cord has two functions: (a) carrying sensory information to the brain and motor output to nerves and (b) mediating spinal reflexes. Spinal nerves send information from peripheral sensory receptors to the spinal cord and information from the spinal cord to effectors (muscles and glands).

A. PROTECTIVE STRUCTURES AND SPINAL MENINGES

The **spinal cord** is protected by the bony vertebrae, adipose tissue, spinal meninges, and cerebrospinal fluid. Three **meninges** (meninx, singular) or connective tissue membranes cover the spinal cord and are continuous with the cranial meninges that protect the brain. **Dura mater,** the outer meninx, is a tough, single-layered membrane that is deep to the adipose tissue in the epidural space and superficial to the spider web-like **arachnoid mater.** The inner meninx, the **pia mater** is delicate and hugs the spinal cord. Between the pia and arachnoid mater is the subarachnoid space that contains cerebrospinal fluid.

- arachnoid mater with web-like projection
- dura mater
- adipose tissue in epidural space
- pia mater
- subarachnoid space

1 _____

2 _____

3 _____

4 _____

5 _____

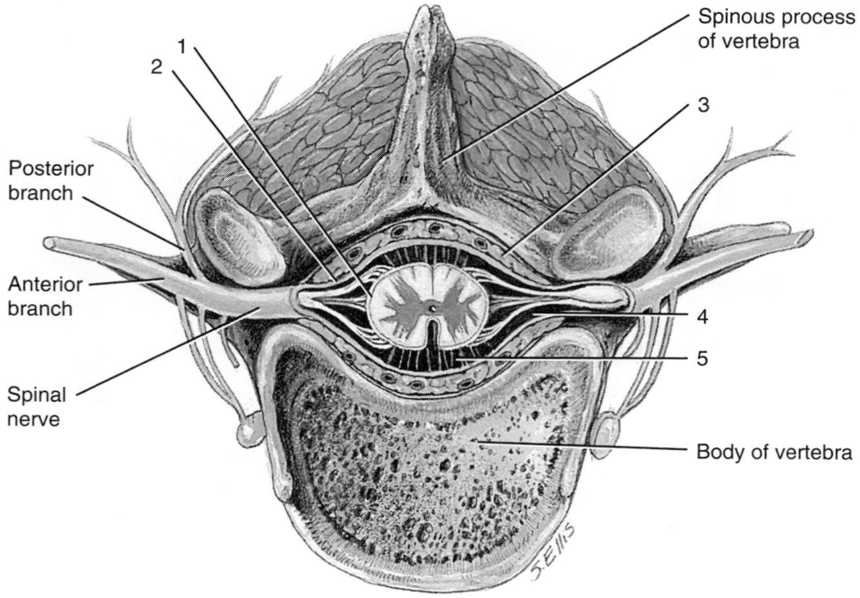

FIGURE 14.1   Spinal cord transverse section showing meninges.

ACTIVITY 1 PROTECTIVE STRUCTURES AND SPINAL MENINGES

1 Label the structures listed in Figure 14.1.

2 Color the meninges in contrasting colors.

3 Identify these structures on a spinal cord model or chart.

4 Pronounce the terms as you point to them.

B. EXTERNAL FEATURES OF THE SPINAL CORD AND SPINAL NERVES

The long, cylindrical spinal cord has spinal nerves that emerge from it and exit through **intervertebral foramina.** Spinal nerves are named for the vertebral region and level from which they arise. The spinal cord begins at the foramen magnum and ends inferiorly between vertebra levels L1 and L2 as the **conus medullaris.** Nerves arising from the inferior portion of the spinal cord continue inferiorly as a group called the **cauda equina** (*cauda* = tail; *equin-* = horse), or "horse's tail." An extension of the pia

mater continues past the conus medullaris as the **filum terminale** (*filum* = filament; *termin-* = terminal) and connects the spinal cord to the coccyx. Of the 31 pairs of spinal nerves, there are **8 cervical, 12 thoracic, 5 lumbar, 5 sacral,** and **1 coccygeal.** Distal to where a spinal nerve passes through its intervertebral foramen, the spinal nerve branches (Figure 14.1). Anterior branches from some spinal nerves form a braided network or **plexus** (*plexus* = braid) before they innervate body structures. This occurs in four regions of the body where these networks form the **cervical, brachial, lumbar,** and **sacral plexuses.** The **thoracic** (intercostal) spinal nerves (T2–T12) do not participate in forming plexuses.

ACTIVITY 2 EXTERNAL FEATURES OF THE SPINAL CORD AND SPINAL NERVES

1 Label the structures listed in Figure 14.2.

2 Color the plexuses in contrasting colors.

3 Identify these structures on a spinal cord model or chart.

4 Pronounce the terms as you point to them.

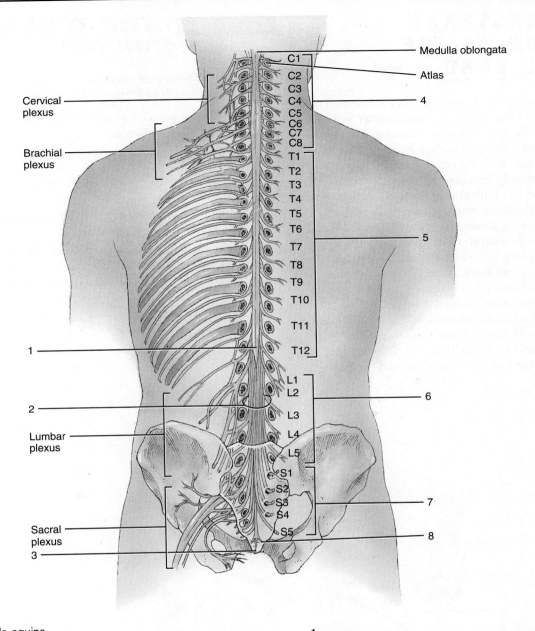

Medulla oblongata

Atlas

4

Cervical plexus

Brachial plexus

5

1

2

Lumbar plexus

6

Sacral plexus

3

7

8

C1
C2
C3
C4
C5
C6
C7
C8
T1
T2
T3
T4
T5
T6
T7
T8
T9
T10
T11
T12
L1
L2
L3
L4
L5
S1
S2
S3
S4
S5

- cauda equina
- cervical nerves
- coccygeal nerve (cox-sih-GEAL)
- conus medullaris
- filum terminale
- lumbar nerves
- sacral nerves
- thoracic nerves

1 _____

2 _____

3 _____

4 _____

5 _____

6 _____

7 _____

8 _____

FIGURE 14.2   Posterior view of the four spinal plexuses.

C. TRANSVERSE SECTION OF THE SPINAL CORD

The most obvious parts of the spinal cord in cross-section are the gray and white matter. The **gray matter** looks like a butterfly or a modified H and is more centrally located. The gray matter consists of nerve cell bodies and dendrites. **White matter** surrounds the gray matter and is composed of white, myelinated fibers (axons) that are either sensory or motor fibers. The **central canal** is in the center of the gray matter and contains cerebrospinal fluid.

Spinal nerves are formed from the **posterior (dorsal) root** and **anterior (ventral) root.** Roots are collections of axons that are going to and leaving the spinal cord. The posterior (dorsal) root carries sensory fibers, whereas the anterior (ventral) root carries motor fibers. Because the sensory and motor roots merge to form the spinal nerve, these nerves are called **mixed nerves.** Near the spinal cord, there is a bulge in the posterior (dorsal) root called the **posterior (dorsal) root ganglion.** The posterior (dorsal) root ganglion consists of somatic sensory neuron cell bodies.

ACTIVITY 3 TRANSVERSE SECTION OF SPINAL CORD

1 Label the spinal cord structures in Figure 14.3 and the photomicrograph in Figure 14.4.

2 Identify these parts on a model or chart of a transverse section of the spinal cord.

3 Pronounce these terms as you point to them.

4 Examine a prepared slide of a transverse section of the spinal cord.

- Using the low-power objective lens, identify the structures listed in Figure 14.3.

- Using the high-power objective lens, observe white matter and gray matter.

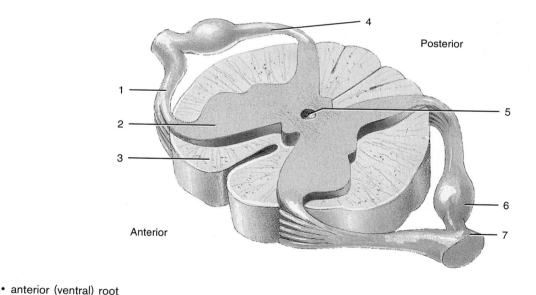

Posterior

Anterior

- anterior (ventral) root
- central canal
- gray matter
- posterior (dorsal) root
- posterior (dorsal) root ganglion
- spinal nerve
- white matter

1 _____   5 _____

2 _____   6 _____

3 _____   7 _____

4 _____

FIGURE 14.3 **Transverse section of spinal cord with areas of gray and white matter.**

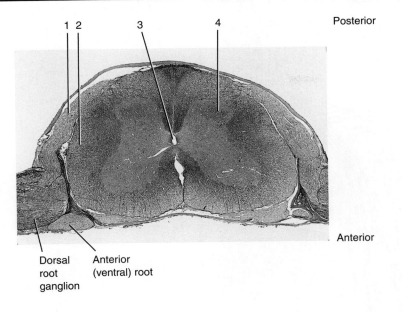

Posterior

Anterior

Dorsal root ganglion

Anterior (ventral) root

- central canal
- gray matter
- posterior (dorsal) root
- white matter

1 _____   3 _____

2 _____   4 _____

FIGURE 14.4 **Photomicrograph of spinal cord.**

D. MAJOR NERVES FROM THE FOUR SPINAL NERVE PLEXUSES

The **cervical plexus** is formed from the anterior (ventral) branch of C1 to C5 on both the right and left sides of the spinal cord. An important paired nerve from this plexus is the **phrenic nerve** C3 to C5, that innervates the diaphragm and is important for breathing. Remember the saying: *Cervical nerves 3, 4, and 5, keep the diaphragm alive.* Other cervical nerves mainly supply the scalp, neck, shoulders, and chest.

The **brachial plexus** is formed from the anterior (ventral) branch of C5 to T1. This plexus serves the shoulders and upper limbs. The main nerves that arise from the brachial plexus are the **axillary, median, musculocutaneous** (*musculo-* = muscle; *cutan-* = skin), **radial,** and **ulnar.**

The **lumbar plexus** is made up of anterior (ventral) branch from L1 to L4. The lumbar plexus supplies the skin and muscles of the abdominal wall, external genitalia, and part of the lower limbs. The major nerves are the **femoral** and **obturator nerves.**

The **sacral plexus** is formed from the anterior (ventral) branch from L4 to S4. This plexus supplies the buttocks, perineum, and most of the lower limbs. Its major nerves are the **pudendal** and **sciatic.** The sciatic becomes the **tibial** and **common fibular (common peroneal)** nerves in the leg.

ACTIVITY 4 MAJOR NERVES FROM THE CERVICAL, THORACIC, LUMBAR, AND SACRAL PLEXUSES

1 Color the nerves in Figures 14.5 and 14.6 in contrasting colors.

2 Identify the nerves in Figures 14.5 and 14.6 on models or charts.

3 Pronounce the terms as you point to them.

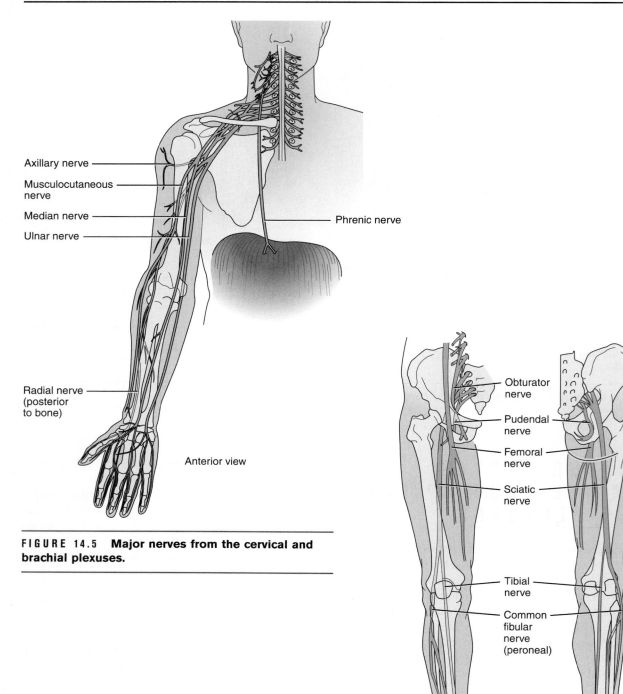

Axillary nerve

Musculocutaneous nerve

Median nerve

Ulnar nerve

Phrenic nerve

Radial nerve (posterior to bone)

Anterior view

FIGURE 14.5   Major nerves from the cervical and brachial plexuses.

Obturator nerve

Pudendal nerve

Femoral nerve

Sciatic nerve

Tibial nerve

Common fibular nerve (peroneal)

Anterior

Posterior

FIGURE 14.6   Major nerves of the lumbar and sacral plexuses.

E. DISSECTION OF THE SPINAL CORD

Use either a preserved cow or sheep spinal cord or a fresh spinal cord from a butcher. Remember, if the specimen is preserved, it will be firmer and look different from a fresh specimen.

> **LAB SAFETY:** *Use safety glasses and remember to properly dispose of any removed waste tissue as directed by your instructor. Always wash your dissection tools, tray, lab surfaces, and hands when you are finished.*

ACTIVITY 5 SPINAL CORD DISSECTION

1 Observe the posterior structures. Refer to Figure 14.7.

- After putting on your gloves, place the spinal cord in the dissection pan.
- Use a blunt probe or forceps to separate the **spinal meninges.**
- Use a pointed dissection probe or pin to detach the **pia mater** from the spinal cord.
- Peel back the meninges to uncover the **posterior (dorsal) roots** and **anterior (ventral) roots** of the spinal nerve.
- Identify the **posterior (dorsal) root ganglion.**

2 Observe the transverse section structures.

- Observe a demonstration of a 1 × 4-inch to 1 × 2-inch section from the spinal cord.
- Identify the **gray matter, white matter,** and **central canal.**

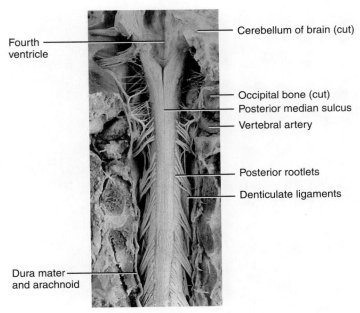

SUPERIOR

Fourth ventricle — Cerebellum of brain (cut) — Occipital bone (cut) — Posterior median sulcus — Vertebral artery — Posterior rootlets — Denticulate ligaments — Dura mater and arachnoid

INFERIOR

FIGURE 14.7 Photograph of posterior view of spinal cord with meninges and spinal nerve roots.

REVIEWING YOUR KNOWLEDGE

A. Meninges

Match the terms with the correct description.

a. arachnoid mater
b. dura mater
c. epidural space
d. pia mater
e. subarachnoid space

_____ **1.** web-like middle meninx

_____ **2.** tough outer meninx

_____ **3.** thin meninx intimate with spinal cord

_____ **4.** contains cerebrospinal fluid

_____ **5.** superficial to the dura mater; filled with adipose tissue

B. Identification of Spinal Cord Structures

Label the spinal cord structures in Figure 14.8.

1. _____

2. _____

3. _____

4. _____

5. _____

6. _____

7. _____

8. _____

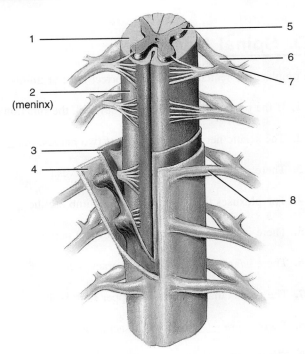

FIGURE 14.8 Anterior view of spinal cord with meninges and gray and white matter.

C. Spinal Cord Structures

Write the term next to the correct description.

anterior (ventral) root
cauda equina
central canal
conus medullaris
filum terminale

gray matter
posterior (dorsal) root
posterior (dorsal) root ganglion
white matter

_____ **1.** Contains neuron cell bodies and unmyelinated processes.

_____ **2.** Sensory branch entering spinal cord.

_____ **3.** Tapered end of spinal cord.

_____ **4.** Motor branch exiting spinal cord.

_____ **5.** Contains sensory neuron cell bodies.

_____ **6.** Collection of spinal nerves that arise from inferior end of spinal cord.

_____ **7.** Composed of myelinated axons.

_____ **8.** Small central passageway of spinal cord that contains cerebrospinal fluid.

_____ **9.** Extension of pia mater that attaches spinal cord to coccyx.

D. Spinal Nerves

Complete the sentences with the correct term or number.

1. If the posterior (dorsal) root is severed, the functional loss is _____ (motor or sensory).

2. The sciatic nerve is comprised of two nerves, the _____ and the _____.

3. There is/are _____ pair(s) of spinal nerves.

4. The plexuses that supply the lower limb are the _____ and _____.

5. There is/are _____ pair(s) of coccygeal nerves.

6. There is/are _____ pair(s) of thoracic nerves.

7. The plexus that supplies the upper limb is the _____.

8. There is/are _____ pair(s) of sacral nerves.

9. There is/are _____ pair(s) of lumbar nerves.

10. There is/are _____ pair(s) of cervical nerves.

11. The _____ nerves supply the diaphragm for breathing.

12. The foramen that the spinal nerves use to exit the spinal cord are called the _____ foramen.

SOMATIC REFLEXES

OBJECTIVES

After completing this exercise, you should be able to:

- Identify and describe the five components of a somatic reflex arc on models or charts
- Explain the significance of clinical testing of reflexes

MATERIALS

- reflex hammer (with rubber head)

Reflexes are rapid, involuntary motor responses to an environmental stimulus detected by sensory receptors. A nerve impulse travels from the receptor through a neural **reflex arc** pathway to an effector. If the motor response is contraction of skeletal muscle, the reflex is called a **somatic reflex.** If the motor response involves cardiac muscle, smooth muscle, or glands, the reflex is called an **autonomic (visceral) reflex.** Reflexes mediated by spinal nerves are called **spinal reflexes,** whereas reflexes mediated by cranial nerves are called **cranial reflexes.** Most reflexes help our bodies maintain homeostasis and therefore have a protective function.

A. REFLEX ARCS

There are five basic components of a reflex arc:

1 Sensory receptor. If the stimulus to the sensory receptor is strong enough, an action potential is generated in the sensory neuron.

2 Sensory neuron. The sensory neuron propagates the action potential and synapses with neurons in the spinal cord or brain stem.

3 Integrating center. The integrating center is located within the gray matter of the central nervous system (CNS) and transfers information from the sensory neuron to the motor neuron.

4 Motor neuron. The motor neuron carries the action potential initiated by the integrating center to the effector.

5 Effector. An effector can be skeletal muscle (somatic reflex), cardiac muscle, smooth muscle, or glands (somatic reflex). A **reflex** is the response of the effector to stimulation by the motor neuron of the reflex arc.

ACTIVITY 1 SOMATIC REFLEXES

1 Label the reflex arc in Figure 15.1.

- effector
- integrating centers
- motor neuron
- sensory neuron
- sensory receptor

1 _____   4 _____

2 _____   5 _____

3 _____

FIGURE 15.1 Reflex arc components.

B. REFLEX TESTING

A series of reflex tests are used clinically to evaluate the nervous system and to diagnose an abnormality or dysfunction. The following reflexes are examples of clinical reflex tests.

1. Patellar Reflex (Knee Jerk)

The **patellar reflex** is the extension of the knee that occurs when the patellar tendon is stretched. The components of the **patellar reflex arc** are:

- *Sensory receptors:* Muscle spindles (stretch receptors) are located in the quadriceps femoris muscle group. Tapping the patellar tendon stretches this muscle group and stimulates muscle spindles, initiating nerve impulses in axons of sensory neurons.

- *Sensory neuron:* Sensory axons carry nerve impulses to the integrating center (gray matter) in the spinal cord.

- *Integrating center:* The sensory axon synapses with and initiates nerve impulses in the motor neuron that innervates the quadriceps femoris muscle group.

- *Motor neuron:* Axons of the motor neuron travel in the femoral nerve to the quadriceps femoris muscle group.

- *Effector:* The quadriceps femoris muscle group (agonists) contracts and extends the leg when stimulated.

ACTIVITY 2 TESTING THE PATELLAR REFLEX

1 Test the patellar reflex.

- Choose one member of your group to be the subject. Have the subject sit on the edge of a table or a tall lab chair with the knees off the table or chair and the legs dangling and relaxed.

- Palpate the patella and the tibial tuberosity. The patellar tendon, the insertion of the quadriceps femoris, is between these two structures. To be sure you have found the tendon, have the subject isometrically flex the quadriceps muscles, but not move the leg while you palpate the tendon.

- Gently but firmly tap the patellar tendon with the tapered end of a reflex hammer.

- Although the grading is subjective, grade the extension of the leg. Results: _____

0 = no response (hyporeflexive)
+1 = little response (hyporeflexive)
+2 = normal response
+3 = above normal response
+4 = exaggerated reflex (hyperreflexive)

- Now ask the subject to clasp both hands in front of the chest and isometrically pull in opposite directions. This action leads to an enhancement of spinal reflexes causing a *reinforced* patellar reflex.
- While the subject is pulling, strike the patellar tendon again and observe the distance of leg movement this time.
- Reinforced patellar reflex results: _____

2. Biceps Reflex (Biceps Jerk)

The **biceps reflex** is the contraction of the biceps brachii muscle that occurs when the biceps tendon is stretched. Although the biceps brachii flexes the forearm, in this test you may only see the contraction or tension in the biceps brachii muscle because the forearm is resting on the table.

ACTIVITY 3 TESTING THE BICEPS REFLEX (BICEPS JERK)

1 Test the biceps reflex.
- Have the subject stand with an arm completely relaxed and hanging down at the side.
- Ask the subject to isometrically contract the biceps brachii so you can palpate the tendon in the antecubital fossa.
- After locating the tendon, have the subject relax and place your thumb over the tendon.
- Tap your thumb with the tapered end of the reflex hammer.
- Biceps reflex results: _____
- If the reflex is absent, have the subject clench his or her teeth for a reinforced reflex and try the test again.
- Reinforced biceps reflex results: _____

3. Achilles Reflex (Ankle Jerk)

The response of this reflex is plantar flexion when the Achilles (calcaneous) tendon is tapped with the reflex hammer.

ACTIVITY 4 ACHILLES REFLEX (ANKLE JERK)

1 Test the Achilles reflex.
- Have the subject keep one foot on the floor and place the knee of the other leg on a chair with the foot dangling over the edge.
- Ask the subject to slightly dorsiflex the foot to stretch the tendon a little, but remain relaxed.
- Tap the calcaneal tendon with the tapered end reflex hammer and observe the response of the foot.
- Calcaneal reflex results: _____

4. Plantar Flexion

This test stimulates the cutaneous receptors and involves the brain in addition to the spinal cord. In adults, **plantar flexion** with flexed (curled) toes occurs when the plantar surface of the foot is stroked. A reaction of dorsiflexion with extended flared toes (Babinski sign) is seen in infants because all the nerve fibers are not myelinated. The Babinski sign is abnormal in adults.

ACTIVITY 5 PLANTAR FLEXION

1 Test the plantar flexion reflex.
- Have the subject seated with a foot propped up and relaxed.
- Using the metal end of the reflex hammer, stroke the plantar surface of the foot starting at the heel, extending up the lateral side of the foot, and crossing over to the great toe area.
- Plantar flexion results: _____

REVIEWING YOUR KNOWLEDGE

1. Name the five components of a reflex arc in order.

a. _____

b. _____

c. _____

d. _____

e. _____

2. Define reflex.

3. Describe the difference between a somatic and visceral reflex.

4. Describe the difference between a cranial and spinal reflex.

5. Name the nerve that is tested in the patellar reflex test.

6. Describe the effect that clasping and pulling the hands had on the patellar reflex.

7. Describe the effect that clenching the teeth had on the biceps reflex.

8. Name the muscles that contract for the Achilles tendon reflex.

9. Name the nerve tested in the Achilles reflex and plantar reflex tests.

16

BRAIN AND CRANIAL NERVES

OBJECTIVES

After completing this exercise, you should be able to:

- Identify the major structures of the brain on models or charts and describe their functions

- Name the three meninges and describe their similarities and differences

- Describe the production of cerebrospinal fluid and trace its circulation

- Identify the 12 pairs of cranial nerves by name and Roman numeral on brain models and/or preserved human brains

- State the function of the 12 pairs of cranial nerves

- Compare the anatomy of the sheep brain with that of the human brain

MATERIALS

- human brain models and charts
- ventricular system model or chart
- preserved human brain
- sheep brain dissection: preserved sheep brains, dissecting instruments, and dissection trays
- disposable gloves and safety glasses
- colored pencils

The human brain simultaneously conducts and coordinates a variety of complicated processes. Its expansive development distinguishes it from all other creatures. Our brain, one of the largest organs in the human body, is responsible for our memory, intellect, ideas, and behavior. The brain is the center for cataloging sensory information, integrating this information with previously recorded information, and producing actions based on the results of the information synthesis.

A. MAJOR BRAIN REGIONS

The brain is composed of four main regions: the brain stem, diencephalon (*di-* = through; *encephal-* = brain), cerebellum (*cerebel-* = little brain), and cerebrum (*cerebr-* = brain). The **brain stem** is connected to the superior part of the spinal cord, and the **diencephalon,** which is in the center of the brain, is superior to the brain stem. The **cerebellum** is posterior to the brain stem. The large **cerebrum,** which envelops the diencephalon, is the dominant brain structure.

1. The Brain Stem

The brain stem is comprised of three structures: the medulla oblongata (*medull-* = marrow), the pons, and the midbrain. The **medulla oblongata** is immediately superior to the spinal cord and is the most vital part of the brain because it houses the respiratory and cardiovascular control centers. The respiratory center controls the rate and depth of breathing, and the cardiovascular center is responsible for the rate and force of the heartbeat and blood pressure reflexes. The medulla oblongata also contains sensory and motor tracts which relay information between the spinal cord and higher brain centers.

The **pons** is an expanded structure located superior to the medulla oblongata and anterior to the cerebellum. It has respiratory centers that assist the medulla oblongata in controlling breathing. The pons also relays information to the diencephalon and the cerebellum.

The **midbrain** is a smaller area superior to the pons and inferior to the diencephalon and contains nerve tracts that connect the upper and lower brain areas. The midbrain is best observed from a midsagittal section. It spans from the dorsal to the ventral side of the brain.

2. The Diencephalon

The **diencephalon** (*di-* = two; *-cephalon* = brain) is located in the brain's central area and contains the thalamus and the hypothalamus. The **thalamus** (*thala-* = inner chamber) is composed of paired, egg-shaped bodies centrally located in the diencephalon that comprise approximately 80% of this structure. The thalamus is the brain's "Grand Central relay station," because it is the principal relay station for sensory fibers. These sensory fibers synapse in the thalamus and relay sensory information to a particular area of the cerebral cortex to be interpreted. The thalamus also filters out unnecessary sensory information and is involved in consciousness, emotions, learning, and memory.

The **hypothalamus** (*hypo-* = below) is located below the thalamus. The hypothalamus has important nuclei (nerve cell bodies) that control many body functions and homeostasis. Some of the major functions include integrating and controlling the pituitary gland and hormonal functions, autonomic nervous system functions, emotions and behavior, eating and drinking, and body temperature.

Other structures that are observed in this area are the pineal gland and the pituitary gland. The **pineal gland** is a small endocrine gland located superior and posterior to the thalamus. The pineal gland produces melatonin, a hormone that plays a role in the sleep/awake cycle. The **pituitary gland** looks like a large pea and is attached to the hypothalamus, which controls the pituitary gland.

3. The Cerebellum

The **cerebellum** (*cerebellum* = little brain) is second in size to the cerebrum and is located inferior to it and posterior to the medulla and pons. There are two connected **cerebellar hemispheres.** When cut in sagittal section, gray matter can be observed on the exterior, with the deeper white matter appearing as branches of a tree. The outer surface of the cerebellum has many slender folds that increase its surface area. The cerebellum regulates posture and balance and, in addition, smoothes and coordinates skilled skeletal muscle movements. The cerebellum is connected to the brain stem by nerve tracts.

ACTIVITY 1 THE BRAIN STEM, DIENCEPHALON, AND CEREBELLUM

1 Label the parts of the brain stem, diencephalon, and cerebellum in Figures 16.1 and 16.2.

2 Using contrasting colors, color the brain stem, diencephalon, cerebellum, and cerebrum in the drawing of Figure 16.1.

3 Identify the structures listed on the human brain model, chart, and/or preserved human brain.

4 Pronounce the terms as you point to them.

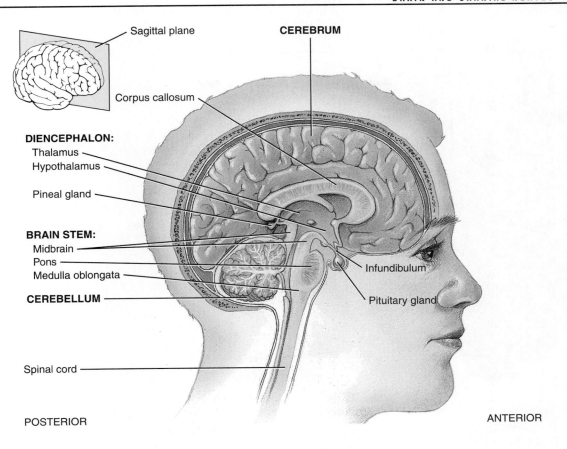

Sagittal plane

CEREBRUM

Corpus callosum

DIENCEPHALON:
Thalamus
Hypothalamus

Pineal gland

BRAIN STEM:
Midbrain
Pons
Medulla oblongata

CEREBELLUM

Infundibulum

Pituitary gland

Spinal cord

POSTERIOR

ANTERIOR

- brain stem
- cerebellum
- cerebrum
- diencephalon
- hypothalamus (hypo-THAL-
  a-mus)
- midbrain
- medulla oblongata
- pineal gland (pi-NEE-al)
- pons
- thalamus

1 _____

2 _____

3 _____

4 _____

5 _____

6 _____

7 _____

8 _____

9 _____

10 _____

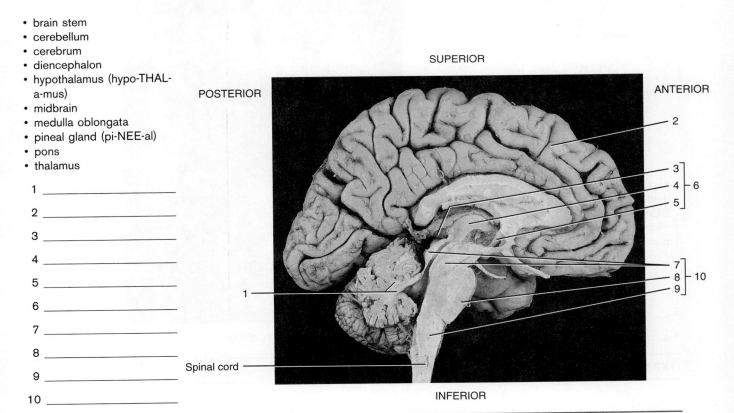

SUPERIOR

POSTERIOR

ANTERIOR

2

3
4 — 6
5

1

7
8 — 10
9

Spinal cord

INFERIOR

FIGURE 16.1   **Major brain regions, midsagittal view.**

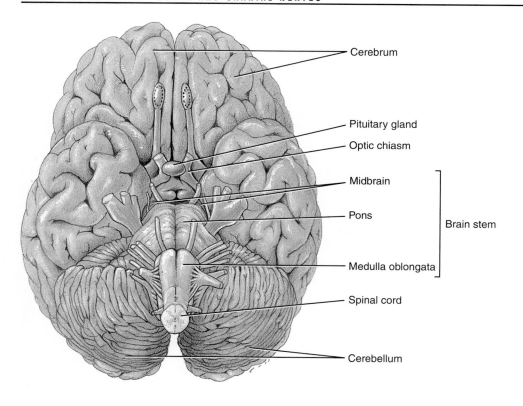

Cerebrum

Pituitary gland

Optic chiasm

Midbrain

Pons

Brain stem

Medulla oblongata

Spinal cord

Cerebellum

Inferior

ANTERIOR

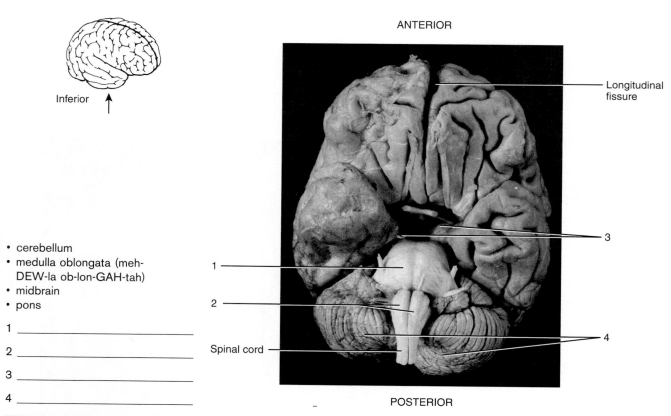

Longitudinal fissure

3

1

2

4

Spinal cord

POSTERIOR

- cerebellum
- medulla oblongata (meh-DEW-la ob-lon-GAH-tah)
- midbrain
- pons

1 _____

2 _____

3 _____

4 _____

FIGURE 16.2 Major brain regions, inferior view.

4. The Cerebrum

The **cerebrum** is made up of right and left **cerebral hemispheres** and is the largest and most complex division of the brain. The cerebrum is superior to and surrounds the diencephalon and part of the brain stem. The cerebrum is the center of higher mental processes such as intelligence, communication, learning and memory, reasoning, and emotions. In addition, it interprets sensory input and initiates skeletal muscle contraction.

There are four lobes composing the exterior of the cerebrum that are mainly named for the overlying cranial bones. These lobes are the **frontal, parietal, occipital,** and **temporal lobes.** An inner lobe, the **insula,** lies deep within the brain and is not visible from the exterior. Other obvious external anatomical features of the cerebrum include the following:

- **sulci** (*sulcus* = furrow; **sulcus,** singular)—shallow grooves between elevations
- **central sulcus**—shallow groove separating the frontal lobe from the parietal lobe
- **gyri** (*gyros* = circle; **gyrus,** singular)—elevations or folds in the cerebral cortex; increase the surface area for neuron cell bodies
- **precentral gyrus**—elevation located just anterior to the central sulcus
- **postcentral gyrus**—elevation located just posterior to the central sulcus
- **longitudinal fissure**—deep groove separating the two cerebral hemispheres at the midline

The **cerebral cortex** (also known as the cortical area) is the superficial **gray matter** on the exterior of the cerebrum and is composed of nerve cell bodies and dendrites. The cerebral cortex integrates sensory information and initiates motor output. **White matter** lies deep to the outer cortex and is comprised mostly of myelinated axons that give it the white appearance. The **corpus callosum** (*corpus* = body; *callosus* = hard), which is readily observable in midsagittal sections of the brain, is a band of myelinated axons that connects the two cerebral hemispheres.

ACTIVITY 2 THE CEREBRUM

1 Label the structures listed in Figures 16.3 and 16.4(a) and (b). Note that the brain is tilted forward in the photograph in Figure 16.3.

2 Using contrasting colors, color the lobes of the brain in Figure 16.3.

3 Identify these features on a brain model or chart.

4 Pronounce the terms as you point to them.

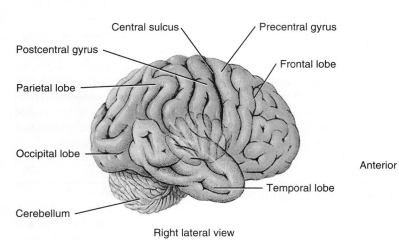

Right lateral view

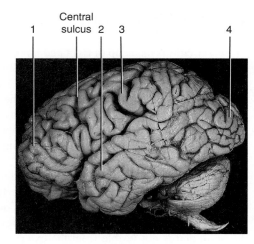

Anterior

Left lateral view
• frontal lobe
• occipital lobe
• parietal lobe
• temporal lobe

Left lateral view

1 _____

2 _____

3 _____

4 _____

FIGURE 16.3 **Lobes of the brain.**

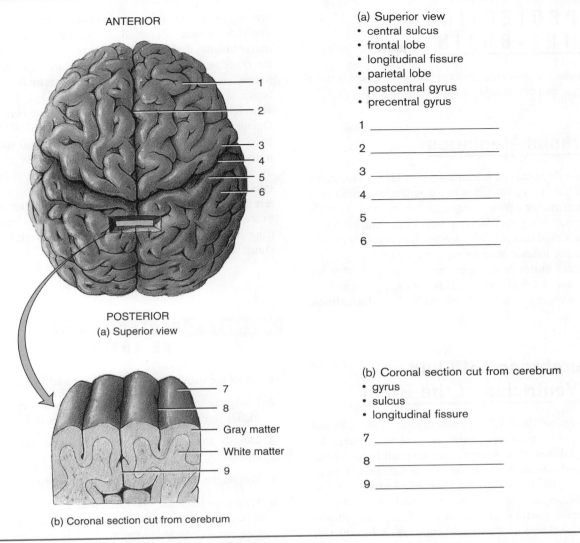

ANTERIOR

POSTERIOR
(a) Superior view

(b) Coronal section cut from cerebrum

ANTERIOR

POSTERIOR

Gray matter

White matter

(a) Superior view
• central sulcus
• frontal lobe
• longitudinal fissure
• parietal lobe
• postcentral gyrus
• precentral gyrus

1 _____

2 _____

3 _____

4 _____

5 _____

6 _____

(b) Coronal section cut from cerebrum
• gyrus
• sulcus
• longitudinal fissure

7 _____

8 _____

9 _____

FIGURE 16.4   External features of the cerebrum.

B. PROTECTION OF THE BRAIN

The brain is protected by cranial bones, the cranial meninges, and cerebrospinal fluid.

1. Cranial Meninges

There are three cranial meninges (connective tissue membranes): the dura mater, arachnoid mater, and pia mater. The **dura mater** (*dura* = hard; *mater* = mother) is the first meninx (singular) located deep to the cranial bones. The **arachnoid** (*arachnoid* = spider-like) **mater** is the second meninx located deep to the dura mater. The **pia** (*pia* = delicate) **mater** is the thin, inner meninx. It hugs and overlays the cerebral cortex, following each gyrus and sulcus. Between the arachnoid and pia is the **subarachnoid** space.

2. Cerebrospinal Fluid and Ventricles of the Brain

Cerebrospinal fluid (CSF) constantly bathes the brain and spinal cord with oxygen, nutrients, and vital chemicals. Although different in content, cerebrospinal fluid is made from blood plasma that leaks out of specialized, tiny blood vessels (capillaries) called the **choroid plexuses** (*choroid* = membrane-like; *plexus* = pleated), and enters into four small brain cavities or ventricles. There are choroid plexuses in the roof of all four ventricles. A **lateral ventricle** is located in each **cerebral hemisphere** with a thin membrane separating the two ventricles anteriorly. An opening in each lateral ventricle connects them to the **third ventricle,** which is medially located between the paired masses of the thalamus and is narrower and smaller than the other ventricles. Connecting the third ventricle to the fourth ventricle is a thin tube, the **cerebral aqueduct** (*aqua* = water; *duct* = way; i.e., waterway). The **fourth ventricle** is located between the **pons** and the **cerebellum.** Openings in the fourth ventricle allow the CSF to flow from the fourth ventricle into the **subarachnoid space** surrounding the brain and the spinal cord. The CSF also flows through the central canal of the spinal cord and back out into the subarachnoid space. Just as CSF is formed and enters the brain ventricles from the bloodstream, it is returned to the blood by reabsorption through the **arachnoid villi** (tiny projections of arachnoid) into the **superior sagittal sinus,** a large vein.

ACTIVITY 3 PROTECTION OF THE BRAIN

1 Label the structures in Figures 16.5 and 16.6(a) and (b).

2 Using contrasting colors, color the meninges and gray matter in Figure 16.5 and the ventricles in Figure 16.6.

3 Identify these structures on a human brain model or chart.

4 Pronounce the terms as you point to them.

5 With your laboratory partners, trace a drop of CSF from its origin in a lateral ventricle through the other ventricles and subarachnoid space, and follow it until it is reabsorbed into the bloodstream.

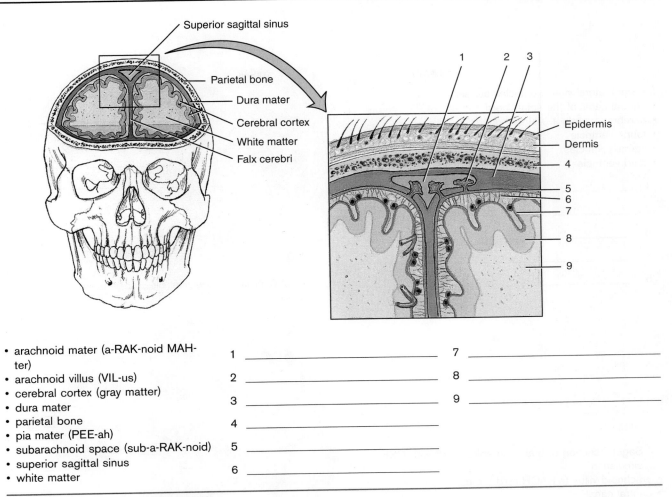

FIGURE 16.5   **Cranial meninges.**

- arachnoid mater (a-RAK-noid MAH-ter)
- arachnoid villus (VIL-us)
- cerebral cortex (gray matter)
- dura mater
- parietal bone
- pia mater (PEE-ah)
- subarachnoid space (sub-a-RAK-noid)
- superior sagittal sinus
- white matter

1 _____   7 _____

2 _____   8 _____

3 _____   9 _____

4 _____

5 _____

6 _____

(a) Right lateral view (ventricles superimposed)
• central canal of the spinal cord
• cerebral aqueduct (AH-que-duct)
• fourth ventricle
• lateral ventricles
• third ventricle

1 _____

2 _____

3 _____

4 _____

5 _____

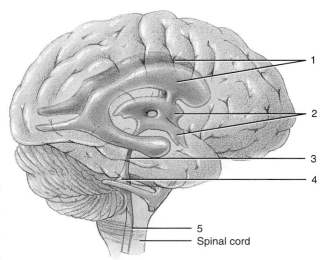

(a) Right lateral view (ventricles superimposed)

(b) Sagittal section of brain and spinal cord with CSF
circulation
• arachnoid villus (a-RACH-noid VIL-us)
• central canal
• cerebral aqueduct
• choroid plexus (CORE-oid PLEX-us)
• fourth ventricle
• lateral ventricle
• subarachnoid space (sub-ah-RAK-noid)
• superior sagittal sinus
• third ventricle

6 _____

7 _____

8 _____

9 _____

10 _____

11 _____

12 _____

13 _____

14 _____

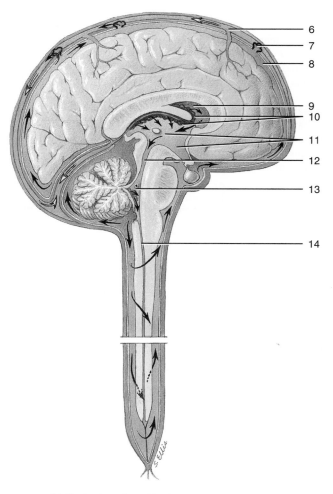

(b) Sagittal section of brain and spinal cord

FIGURE 16.6   Ventricles of the brain and cerebrospinal fluid circulation.

C. NAME AND LOCATION OF CRANIAL NERVES

Cranial nerves are categorized as a part of the peripheral nervous system. The 12 pairs of cranial nerves are numbered sequentially from anterior to posterior. These nerves exit from the brain and pass through foramina to reach their destinations. The nerve's name sometimes indicates the structure it innervates or its function. The names of the cranial nerves I to XII are in order as follows: **olfactory** (*olfactus* = sense of smell), **optic, oculomotor** (*oculo-* = eye; *-motor* = mover), **trochlear** (*trochlea* = pulley), **trigeminal** (*tri* = three; *gemini* = twins), **abducens** (*ab-* = away; *-ducens* = to lead), **facial, vestibulocochlear** (*vestibulo-* = vestibule of ear/equilibrium; *cochlear* = cochlea of ear/hearing), **glossopharyngeal** (*glosso-* = tongue; *-pharyngeal* = throat), **vagus** (*vagus* = wandering), **accessory,** and **hypoglossal** (*hypo-* = below; *-glossal* = tongue). The vagus nerve has the distinction of being different from the other 11 cranial nerves in its distribution. As its word derivative states, it is the "wandering nerve" that branches extensively and innervates the viscera of the thoracic and abdominopelvic cavities.

You may want to use the following mnemonic device to help remember the names of the cranial nerves in order: "**O**n **O**ld **O**lympus' **T**owering **T**op **A** **F**riendly **V**iking **G**rew **V**ines **A**nd **H**ops." The bold first letter of each word matches with the first letter in the name of the cranial nerve.

The olfactory nerves are very small and are not seen in the figures in this exercise. These nerves synapse onto olfactory bulbs which can be observed and are marked.

Not all cranial nerves function in the same way. Some cranial nerves are primarily special sensory in function (**S**), some are primarily motor in function (**M**), and some are mixed nerves with both sensory and motor fibers (**B**) (Table 16.1). A mnemonic device for recalling the function of cranial nerves is: "**S**ome **S**ay **M**arry **M**oney **B**ut **M**y **B**rother **S**ays **B**ad **B**usiness **M**arry **M**oney." Motor nerves also have sensory proprioceptors located in the muscles they innervate, but their main function is motor. Cranial nerve VIII has been shown to have some motor activity, but its main function is sensory. Two of the cranial nerves have branches. The trigeminal nerve (V) has three branches: ophthalmic, maxillary, and mandibular. The accessory nerve (XI) has a cranial portion and a spinal portion.

TABLE 16.1 Cranial Nerve Function, Distribution, and Action

NUMBER AND NAME	DISTRIBUTION	ACTION
I Olfactory	Nasal mucosa	Smell
II Optic	Eye	Vision
III Oculomotor	Levator palpebrae superioris; four extrinsic eye muscles (inferior oblique, superior rectus, medial rectus, inferior rectus); ciliary muscle (intrinsic eye muscle); iris muscles of eye (intrinsic eye muscles)	Movement of eyelid; movement of eyeball; accommodation of lens; pupillary constriction
IV Trochlear	Superior oblique muscle of eyeball	Movement of eyeball
V Trigeminal	Ophthalmic branch Maxillary branch Mandibular branch	Cutaneous sensations from ophthalmic, maxillary, and mandibular areas; chewing
VI Abducens	Lateral rectus muscle of eyeball	Movement of eyeball
VII Facial	Tongue; facial, scalp, and neck muscles; lacrimal glands; salivary glands	Taste; facial expression; secretion of tears; salivation
VIII Vestibulocochlear	Semicircular canals and cochlea of ear	Equilibrium and hearing
IX Glossopharyngeal	Tongue; pharyngeal muscles; parotid gland	Taste; swallowing and speech; secretion of saliva
X Vagus	Pharyngeal muscles and epiglottis; smooth muscles of thorax and gastrointestinal (GI) tract; cardiac muscle; glands of GI tract	Taste and somatic sensation from pharynx and epiglottis; swallowing, coughing, and voice production; smooth muscle contraction of GI tract; slows heart rate; secretion by digestive glands
XI Accessory	Cranial portion – muscles of pharynx, larynx, and soft palate Spinal portion – sternocleidomastoid and trapezius muscles	Swallowing Movement of head and shoulders
XII Hypoglossal	Tongue muscles	Speech and swallowing

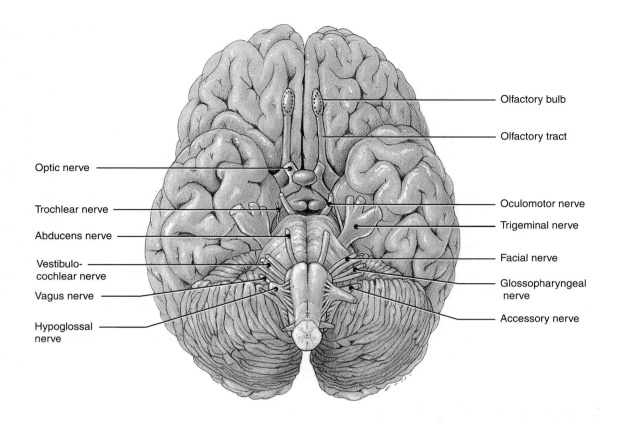

Optic nerve

Trochlear nerve

Abducens nerve

Vestibulo-cochlear nerve

Vagus nerve

Hypoglossal nerve

Olfactory bulb

Olfactory tract

Oculomotor nerve

Trigeminal nerve

Facial nerve

Glossopharyngeal nerve

Accessory nerve

- abducens (ab-DUE-senz)
- accessory
- facial
- glossopharyngeal (gloss-oh-fah-RIN-jeal)
- hypoglossal (hypo-GLOSS-al)
- oculomotor (ok-u-low-MO-tor)
- olfactory (OHL-fac-tory) bulb
- optic
- trigeminal (tri-GEM-i-nal)
- trochlear (TROH-klee-ur)
- vagus (VAY-gus)
- vestibulocochlear (ves-tib-u-lo-COKE-lee-ur)

1 _____

2 _____

3 _____

4 _____

5 _____

6 _____

7 _____

8 _____

9 _____

10 _____

11 _____

12 _____

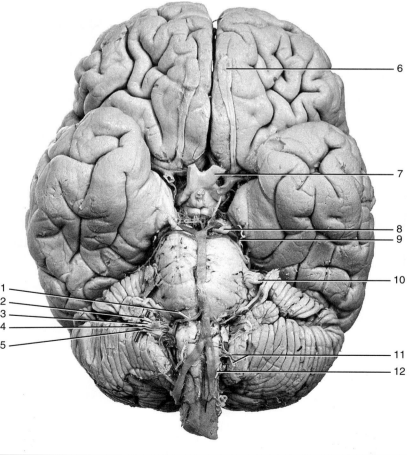

FIGURE 16.7   Cranial nerves of the human brain.

ACTIVITY 4 CRANIAL NERVES

1 Learn the mnemonics for recalling the names and functions of the cranial nerves.

2 Label the cranial nerves on Figure 16.7.

3 Using contrasting colors, color the cranial nerves in Figure 16.7.

4 Identify each cranial nerve on a human brain model or chart.

5 Pronounce the names of the cranial nerves and state their function as you point to them on the brain model or chart.

D. SHEEP BRAIN DISSECTION

SAFETY PRECAUTION:
- *Use disposable gloves, safety glasses, and a lab coat when handling preserved material.*
- *Wash hands after this lab activity.*
- *Follow your instructor's directions for cleaning dissection equipment and table tops.*
- *Consult your instructor about saving or proper disposal of brains.*

Remember that preserved material does not look like a fresh specimen. Usually more detail may be observed in a preserved brain because the tissue is firmer.

ACTIVITY 5 DISSECTION OF SHEEP BRAIN

1 Rinse the sheep brain to remove preservative.

2 Observe meninges and main brain regions.
- Examine the brain to see if the tough, outer **dura mater** is present. Carefully remove the dura mater.

- Now look for the stringy, web-like **arachnoid mater** beneath and adhering to the dura mater. Deep to this membrane is the very thin **pia mater** that follows the contour of the gyri and sulci.
- Compare the sheep brain with the main external regions of the human brain. Identify the **cerebrum, brain stem (medulla and pons),** and **cerebellum.**

3 Identify dorsal structures (Figure 16.8).
- With the dorsal side up, identify the **cerebral hemispheres, gyri, sulci, longitudinal fissure,** and **transverse fissure.**
- Note the four main **lobes** of the brain—**frontal, parietal, occipital,** and **temporal.**
- At the longitudinal fissure, gently separate the two parts and look down between them for the thick band of white fibers, the **corpus callosum.**

4 Identify ventral structures (Figure 16.9).
- Place the sheep brain ventral side up.
- Identify the **olfactory bulb and optic nerve.**
- If present, identify the **pituitary gland.**
- Compare the size of the **brain stem** structures—the **midbrain, pons,** and **medulla oblongata**—with the human brain.

5 Carefully pull the cerebellum away from the cerebrum. Identify the **pineal body.**

6 Identify midsagittal section structures (Figure 16.10).
- Using a sharp knife or scalpel, carefully make a midsagittal section.
- Locate the **brain stem** components: the **medulla oblongata,** the **pons,** and the **midbrain.** Compare this with your human brain model.
- Identify the **cerebellum.** Note the branching white matter in the cerebellum.
- Note the **cerebral aqueduct** and the **fourth ventricle.**
- Identify the **thalamus, corpus collosum,** and **lateral ventricles.**

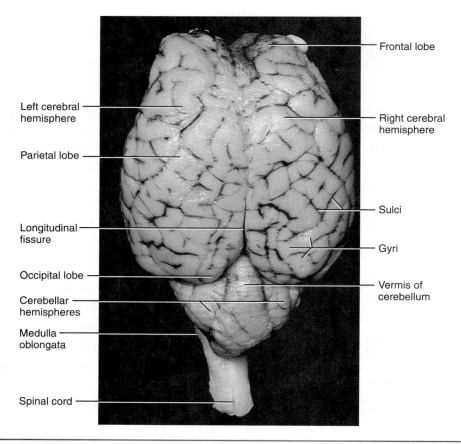

Frontal lobe

Left cerebral hemisphere

Right cerebral hemisphere

Parietal lobe

Sulci

Longitudinal fissure

Gyri

Occipital lobe

Vermis of cerebellum

Cerebellar hemispheres

Medulla oblongata

Spinal cord

FIGURE 16.8   Superior view of the sheep brain.

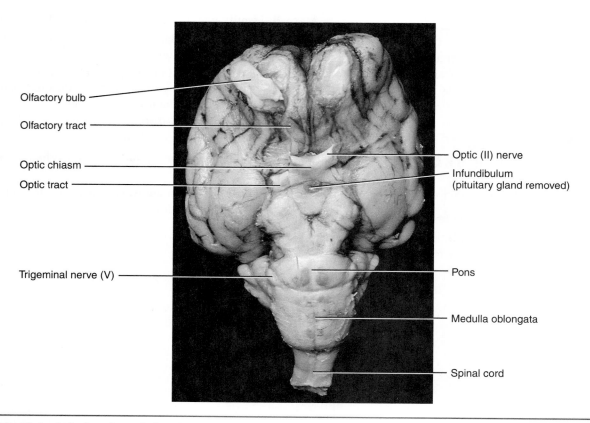

Olfactory bulb

Olfactory tract

Optic chiasm

Optic (II) nerve

Optic tract

Infundibulum (pituitary gland removed)

Trigeminal nerve (V)

Pons

Medulla oblongata

Spinal cord

FIGURE 16.9   Inferior view of the sheep brain.

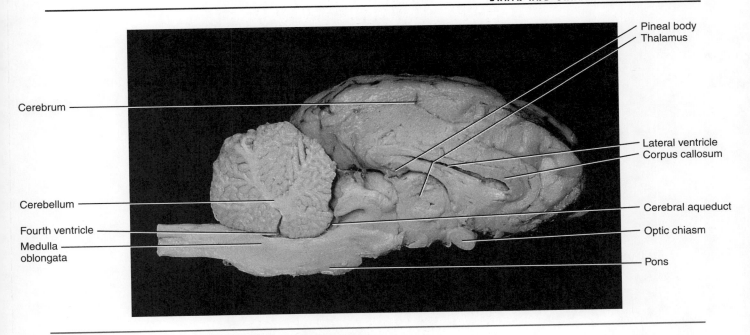

Cerebrum

Cerebellum

Fourth ventricle

Medulla
oblongata

Pineal body
Thalamus

Lateral ventricle
Corpus callosum

Cerebral aqueduct

Optic chiasm

Pons

FIGURE 16.10   Midsagittal section of the sheep brain.

Name ______________________ Section ______________ Date ______________

EXERCISE 16

REVIEWING YOUR KNOWLEDGE

A. Brain Regions and Cranial Nerves

Label the structures on Figure 16.11 in the spaces below.

1. ______________________
2. ______________________
3. ______________________
4. ______________________
5. ______________________
6. ______________________
7. ______________________
8. ______________________
9. ______________________
10. ______________________
11. ______________________
12. ______________________
13. ______________________
14. ______________________
15. ______________________
16. ______________________
17. ______________________
18. ______________________
19. ______________________
20. ______________________

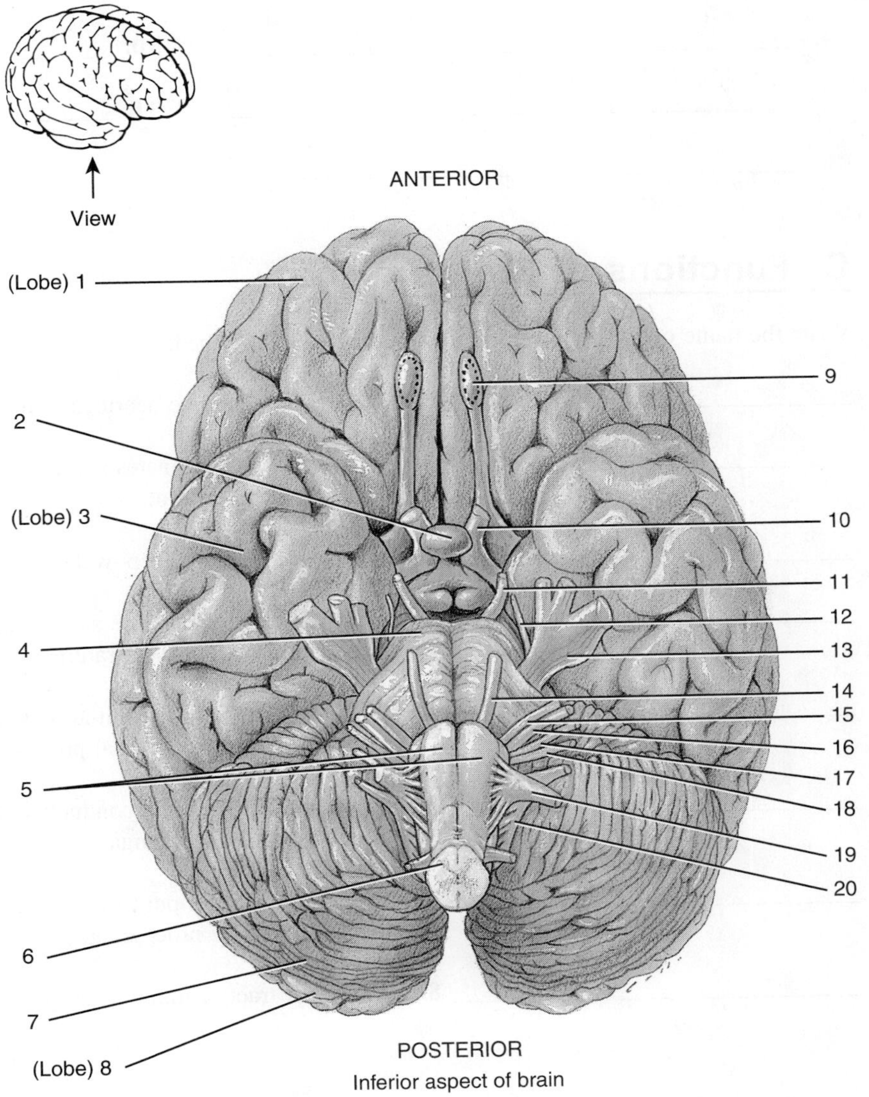

FIGURE 16.11 Ventral view of the human brain.

B. Flow of Cerebrospinal Fluid

Fill in the numbered blanks with the name of the structure that corresponds with the number in the following paragraph.

Fluid for CSF is derived from the bloodstream. The sites of CSF formation are the (1), special tiny capillaries located in the walls of (2), (3), and (4). CSF flows from the two lateral ventricles through an opening into the (5). From here, CSF flows through the (6) into the fourth ventricle. CSF leaves the fourth ventricle and enters the (7) space around the brain, and circulates around the cerebrum and cerebellum. CSF continues to flow into the inner part of the spinal cord by flowing through the tiny (8) of the spinal cord, as well as around the exterior of the spinal cord in the (9) space. Because CSF is continually being made at the rate of about 20 mL/hr, it has to exit back into the bloodstream by being reabsorbed through the (10) that protrude into the dural venous sinuses. The venous sinus that is overlying the brain superiorly is called the (11).

1. _____

2. _____

3. _____

4. _____

5. _____

6. _____

7. _____

8. _____

9. _____

10. _____

11. _____

C. Functions of Brain Regions

Write the name of the brain region for the functions described.

_____ 1. Vital centers regulate heartbeat, breathing, blood pressure, vomiting, coughing.

_____ 2. Smoothes and coordinates skilled skeletal muscle movement; also posture and balance or equilibrium.

_____ 3. Secretes melatonin; sleep–wake cycle.

_____ 4. Controls and integrates the autonomic nervous system; regulates hormones, emotional behavior, temperature, eating, and drinking behavior.

_____ 5. Interprets sensory input, initiates skeletal muscle movements, and is involved in emotional and intellectual processes.

_____ 6. Helps control breathing; conducts impulses to and from the cerebellum, midbrain, and medulla oblongata.

_____ 7. Relays all sensory input to the cerebral cortex; involved in skeletal muscle actions and memory processing.

_____ 8. White fiber tracts communicating between hemispheres.

D. Cranial Nerve Functional Types

Identify if each cranial nerve is mainly sensory, motor, or both.

S = sensory   M = motor   B = both sensory and motor

_____   **1.** Olfactory

_____   **2.** Optic

_____   **3.** Oculomotor

_____   **4.** Trochlear

_____   **5.** Trigeminal

_____   **6.** Abducens

_____   **7.** Facial

_____   **8.** Vestibulocochlear

_____   **9.** Glossopharyngeal

_____   **10.** Vagus

_____   **11.** Accessory

_____   **12.** Hypoglossal

E. Functions of the Cranial Nerves

Write the name of the cranial nerve for the functions described.

_____ 1. Motor nerve to the tongue for speech and swallowing.

_____ 2. Motor nerve to the lateral rectus eye muscle.

_____ 3. Receives sensory information from the nose.

_____ 4. Motor nerve to the superior oblique muscle.

_____ 5. Receives sensory information from the eye for vision.

_____ 6. Motor nerve to the medial, superior, and inferior rectus muscles and inferior oblique muscle.

_____ 7. Transmits cutaneous sensory information from the face and motor information to muscles of mastication.

_____ 8. Motor nerve to the muscles of the neck that move the head.

_____ 9. Mixed nerve of the tongue and pharynx that receives taste information and acts in swallowing.

_____ 10. Receives sensory information from and transmits motor information to the visceral organs.

_____ 11. Receives taste information from the tongue and sends motor information to the facial expression muscles.

_____ 12. Receives sensory information from the ear for hearing and equilibrium.

17

SPECIAL SENSES: THE EYE

OBJECTIVES

After completing this exercise, you should be able to:

- Identify the accessory structures of the eye and the structures of the eyeball on models or charts
- Perform visual tests that are used to determine distance visual acuity, near point of vision, and astigmatism

MATERIALS

- eye models or charts
- dissection equipment and trays, cow eyes, and gloves
- Snellen eye chart

A. STRUCTURE OF THE EYE

1. Accessory Eye Structures

Accessory structures of the eye include the eyebrows, eyelids, eyelashes, lacrimal apparatus, and the extrinsic eye muscles. The **lacrimal apparatus** (*lacrim-* = tears) is a group of structures involved in producing and draining tears. The **lacrimal gland** produces and secretes tears onto the eye surface, and the **lacrimal canals** drain the tears from the eyes. The enlarged **nasolacrimal duct** receives tears from the lacrimal canals and drains the tears into the nasal cavity. The **extrinsic eye muscles** are six skeletal muscles that insert on the exterior of the eyeball to move the eyeball in all directions.

2. Structure of the Eyeball

The wall of the eyeball has three layers: the outer fibrous tunic, the middle vascular tunic, and the inner retina. The **fibrous tunic** is composed of the cornea and sclera. The **cornea** is the transparent anterior portion that covers the iris and pupil, and the **sclera** (*scler-* = hard) is the tough, white part of the eye that forms the majority of the eyeball. The **conjunctiva** is the thin, protective mucous membrane that covers the anterior eye and folds to cover the inner eyelid.

The middle **vascular tunic** is composed of the iris, ciliary body, and choroid. The **iris** is the most anterior portion of the vascular tunic. It is made of circular smooth muscle and controls the pupil size. The **pupil** is the opening in the middle of the iris that allows light to enter the eyeball and changes size in response to the intensity of light. The **ciliary body** begins posterior to the iris at the junction of the cornea and sclera and consists of the ciliary muscle and ciliary processes. The ciliary muscle is a smooth muscle that contracts to control the shape of the lens. The ciliary processes contain capillaries that secrete aqueous humor, the fluid in the anterior chamber of the eyeball. The **choroid** is the most posterior part of the vascular tunic and lines most of the interior of the sclera. It contains many blood vessels that nourish the retina.

The **retina** is the inner coat that starts at the ciliary muscle and continues posteriorly, lining the choroid. The retina contains photoreceptors (rods and cones) that are sensitive to light and initiate nerve impulses that are sent to the brain for vision. The **central fovea** (*fovea* = pit) is located in the center of the **macula lutea** (*macula* = spot; *lutea* = yellow) at the posterior of the eyeball. The central fovea contains the highest concentration of cones. Images focused on the central fovea are sharper (visual acuity) than images focused on other areas of the retina. Rods and cones synapse with other neurons, and the axons of these neurons converge at the posterior–medial part of the eyeball to form the **optic (II) nerve.** The area where axons meet to form the optic nerve is called the **blind spot,** since no photoreceptors are found there.

3. Interior of the Eyeball

The interior of the eyeball contains the lens, anterior cavity, and vitreous chamber. The **lens** divides the interior of the eyeball into an anterior cavity and a vitreous chamber. The **anterior cavity** is a space between the cornea and the lens that is filled with watery **aqueous humor** (*aqua* = water; *humor* = moist). The **scleral venous sinus** (canal of Schlemm) is an opening found at the junction of the cornea and sclera that drains aqueous humor back into the bloodstream. The **vitreous chamber** is the larger, posterior cavity located between the lens and the retina. This cavity is filled with a gel-like substance called the **vitreous body** (humor) that holds the retina flat against the choroid.

ACTIVITY 1 STRUCTURE OF THE EYE

1 Label the structures of the lacrimal apparatus in Figure 17.1.

2 Label the structures of the eyeball in Figure 17.2.

3 Using contrasting colors, color the lacrimal apparatus in Figure 17.1 and the eye structures in Figure 17.2.

4 Identify these structures on an eye model or chart.

5 Pronounce the terms as you point to them.

CLINICAL NOTE: Glaucoma *is caused by an increase in pressure within the eye called intraocular pressure. Blockage of the scleral venous sinus prevents drainage of aqueous humor, increasing the amount in the anterior cavity that causes increased intraocular pressure.* **Macular Degeneration** *is a progressive deterioration of the macula lutea and sight diminishes as cones are destroyed.*

• lacrimal canals (LAK-rih-mal)
• lacrimal gland
• nasolacrimal duct

1 _____

2 _____

3 _____

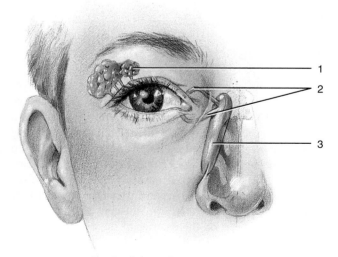

Lacrimal apparatus

FIGURE 17.1   Accessory structures of the eye.

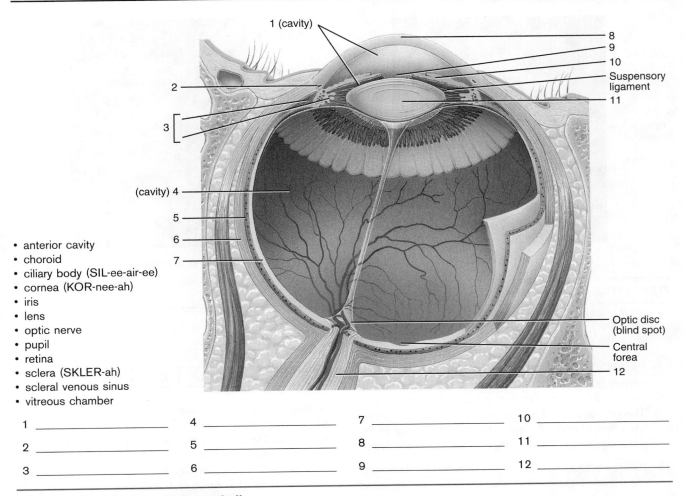

1 (cavity)
8
9
10
Suspensory ligament
11
2
3
(cavity) 4
5
6
7

- anterior cavity
- choroid
- ciliary body (SIL-ee-air-ee)
- cornea (KOR-nee-ah)
- iris
- lens
- optic nerve
- pupil
- retina
- sclera (SKLER-ah)
- scleral venous sinus
- vitreous chamber

Optic disc (blind spot)
Central forea
12

1 _____	4 _____	7 _____	10 _____
2 _____	5 _____	8 _____	11 _____
3 _____	6 _____	9 _____	12 _____

FIGURE 17.2   Structure of the eyeball.

ACTIVITY 2 DISSECTION OF COW EYE

LAB SAFETY: *Use safety glasses and remember to properly dispose of any removed waste tissue as directed by your instructor. Always clean your dissection tools, tray, and lab surfaces as directed.*

1 Obtain a laboratory tray, blunt probe, scalpel, scissors, and disposable gloves.

2 Using gloves, obtain a cow eye and rinse it to remove excess preservative.

3 The posterior portion of the eye may be encased in adipose tissue. Carefully remove the adipose tissue protecting the optic nerve.

4 Identify the external eye structures: **cornea, sclera, optic nerve,** and **extrinsic eye muscles** (refer to Figure 17.3). The preservative causes the cornea to change from transparent and smooth to opaque and wrinkled.

5 Using the point of a scalpel, punch an opening $1/4$-inch posterior to the cornea through the very tough sclera. Be careful not to squeeze the eyeball too tightly or liquid may squirt out. Use scissors to cut an incision all the way around the eyeball, separating it into two parts.

6 Carefully separate the anterior and posterior parts of the eyeball so the **vitreous body** remains in the posterior part of the eyeball and the **lens** in the anterior part. Carefully remove the lens.

7 Using Figure 17.3, identify the **pupil, iris,** and **ciliary muscle** in the anterior portion of the eyeball.

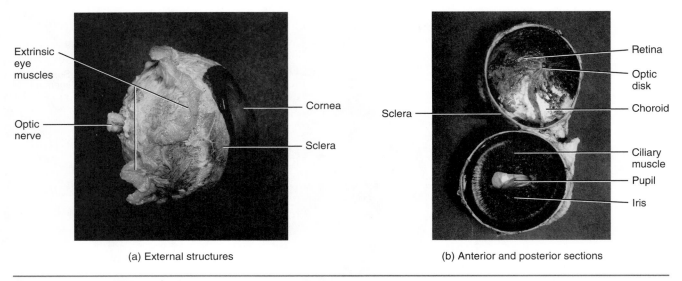

(a) External structures

(b) Anterior and posterior sections

FIGURE 17.3   **Cow eye dissection.**

8   Carefully remove the vitreous body from the posterior portion of the eyeball. Refer to Figure 17.3 to identify the structures in the posterior portion of the eyeball. The **retina,** a thin beige layer, will probably separate from the choroid. The only point of attachment of the retina to the wall of the eyeball is at the **optic disc.**

9   Pull the retina away from the **choroid,** the dark middle layer of the eyeball that contains the pigment melanin. The choroid of the cow's eye has an iridescent reflecting surface that is not found in humans. This irridescent surface reflects light within the eye and enables animals to see in low light conditions.

10 Separate the choroid from the **sclera.** Observe how the three coats or tunics—the **retina, choroid,** and **sclera**—form the wall of the eyeball.

11 Discard the eye as instructed by your instructor.

12 Clean your dissecting tray and equipment, and return them to their proper places.

B. VISUAL TESTS

Visual acuity tests measure the ability of the lens to focus light reflected from an object onto the central fovea of the retina. The lens can **accommodate** or change shape to bend light rays to focus them on the central fovea. At 20 feet, light rays from an object are nearly parallel and do not have to bend as much to focus on the central fovea. At this distance, the lens is flattened and the refractive power (ability to bend light rays) of the lens is lowest. To observe objects closer than 20 feet, the lens must change shape or accommodate to focus light rays on the central fovea. The lens bulges to increase refractive power. Individuals who have normal distance vision and near vision are **emmetropic,** individuals who have normal distance vision but blurry near vision are **hyperopic** (farsighted), and individuals who have blurry distance vision but normal near vision are **myopic** (nearsighted).

As we age, the ability of the lens to accommodate diminishes and the ability to focus on very close objects decreases, a condition called **presbyopia** (*presby-* = old). The **near point of vision** is the closest distance that a person can focus on an object. The average near point of vision is 10 cm for a young adult, 20 cm for an adult in their 40s, and 80 cm for someone in their 60s.

Astigmatism is caused by irregularities in the curvature of the cornea or lens. This causes parts of an image to be blurry.

ACTIVITY 3 VISUAL TESTS

1 **Distance visual acuity** is measured using a **Snellen eye chart** (provided by your instructor). *If you wear eyeglasses or contact lenses, remove them to determine visual acuity without correction or wear them to determine visual acuity with correction.*

- Choose a subject and have him or her stand 20 feet from the Snellen eye chart that is placed in a well-lighted area. Cover left eye with hand.

- Have subject read the smallest line of letters possible without squinting. If the subject can correctly read half of the letters or more, then ask the subject to read the letters on the next smaller line.

- Record the number of the line with the smallest-sized letters read with half or greater accuracy in Table 17.1.

- Cover the right eye and repeat the procedure.

- A value of 20/20 indicates that the subject has normal vision. A value of 20/40 indicates that the subject sees at 20 feet what a person who has normal vision sees at 40 feet. This is not as good as normal vision. A value of 20/15 indicates that the subject sees at 20 feet what a person with normal vision sees at 15 feet. This is better than normal vision.

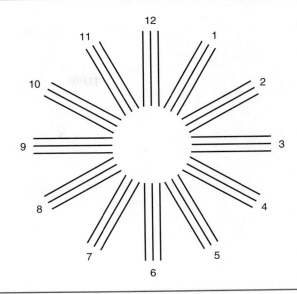

FIGURE 17.4   Astigmatism chart.

2 Measure the subject's **near point of vision** and compare it with the average for his or her age: (young adult = 10 cm; adult in their 40s = 20 cm; adult in their 60s = 80 cm).

- Have the subject hold a textbook 14 inches from his or her face.

- Have the subject cover the left eye and read the words.

- Instruct the subject to slowly move the textbook closer to his or her face until the letters are blurry.

- Measure the distance form the book to the subject's eyes in centimeters.

- Record the value in Table 17.1.

3 Determine whether you have **astigmatism.**

- Remove corrective lenses.

- Cover the right eye and look at the center of the astigmatism chart in Figure 17.4.

- If all the radiating lines are equally sharp and dark, you do not have astigmatism. However, if some lines are lighter or less distinct than others, you have astigmatism.

- Cover the left eye and repeat the procedure.

- Record the results in Table 17.1.

TABLE 17.1 Visual Test Results

TEST	RESULTS
Distance visual acuity test (corrected or uncorrected)	Left eye Right eye
Near point of vision (corrected or uncorrected)	
Astigmatism	Left eye Right eye

REVIEWING YOUR KNOWLEDGE

A. Eye Structures

Choose a structure that applies to each statement. More than one structure may apply to a statement and a structure may be used more than once.

a. anterior cavity
b. choroid
c. ciliary body
d. conjunctiva
e. cornea
f. iris
g. lacrimal canals
h. lacrimal gland

i. lens
j. nasolacrimal duct
k. pupil
l. retina
m. sclera
n. scleral venous sinus
o. vitreous chamber

_____ **1.** Structures that are part of vascular tunic.

_____ **2.** Contains photoreceptors.

_____ **3.** Controls the size of the pupil.

_____ **4.** Drains aqueous humor from the anterior chamber.

_____ **5.** Structures that are part of the fibrous tunic.

_____ **6.** Most anterior part of the eyeball.

_____ **7.** Contains vitreous body.

_____ **8.** Changes shape to focus light on retina.

_____ **9.** Location of aqueous humor.

_____ **10.** White, tough, outer layer of eyeball.

_____ **11.** Membrane that covers the anterior eye and folds to cover the eyelid.

_____ **12.** Drains tears into nasal cavity.

_____ **13.** Produces tears.

_____ **14.** Tears from surface of eye drain into here.

B. Visual Acuity Tests

Answer the following questions.

1. Why must a person stand 20 feet from the Snellen eye chart to test distance vision?

2. Define accommodation.

3. A 10-year-old patient's distance visual acuity was tested and determined to be 20/80. Explain what that means.

4. A 60-year-old man has to hold his newspaper at arm's length to read it. What condition does he have?

5. A person with _____ has blurry near vision but normal distance vision.

6. A person with _____ does not see everything within a visual field clearly; some parts are blurry.

7. A person with _____ has normal near vision but blurry distance vision.

SPECIAL SENSES: THE EAR

OBJECTIVES

After completing this exercise, you should be able to:

- Identify the major structures of the external, middle, and internal ear on models or charts
- Perform tests that are used in determining the cause of hearing loss
- Perform tests that exhibit the function of the receptors for dynamic and static equilibrium

MATERIALS

- ear models or charts
- skull, tuning forks, cotton balls
- swivel chair or stool
- colored pencils

A. ANATOMY OF THE EXTERNAL AND MIDDLE EAR

The ear is divided into three regions: the external (outer) ear, the middle ear, and the internal (inner) ear. The **external ear** consists of the auricle, external auditory canal, and tympanic membrane and extends from the auricle to the tympanic membrane. The auricle, the flexible external structure that is commonly called the ear, collects sound waves and directs them toward the external auditory canal. The **external auditory canal** conducts sound waves from the auricle to the tympanic membrane. The **tympanic membrane** (*tympan-* = drum) or eardrum converts sound waves to vibrations that are transferred to middle ear structures.

The **middle ear** is an air-filled cavity within the temporal bone that extends from the tympanic membrane to the oval window. Middle ear structures include the auditory oscicles, auditory tube, and the oval window. **Auditory oscicles** are small bones within the cavity that are connected by synovial joints. These bones transfer vibrations from the tympanic membrane to the oval window. The **malleus** is the outermost bone and is attached to the tympanic membrane. The **incus** is the middle bone and connects to the stapes. The **stapes** is the innermost bone and connects to the incus and oval window. The **oval window** is the membrane-covered opening that separates the middle and inner ear, and transfers vibrations to the inner ear. The **auditory tube** (pharyngotympanic or Eustachian tube) connects the middle ear to the nasopharynx (part of the throat near the nasal cavity), and equalizes the air pressure of the middle ear with atmospheric air.

The **internal ear** is housed within the temporal bone. It consists of cavities within the bone called the **bony labyrinth** that contains a series of connected membranous sacs, the **membranous labyrinth.** The bony labyrinth contains a fluid called **perilymph** that surrounds the membranous labyrinth. **Endolymph** is the fluid within the membranous labyrinth. The bony labyrinth has three main regions: the vestibule, the semicircular canals, and the cochlea. The **vestibule** is the middle area of the bony labyrinth. The membranous labyrinth within the vestibule contains equilibrium receptors. The **semicircular canals** are three bony canals posterior to the vestibule that project posteriorly, laterally, and superiorly from the vestibule; each canal is at right angles to the other two. The **ampulla** is the widened end of each semicircular canal and contains equilibrium receptors. The **cochlea** is the spiral area of the bony labyrinth anterior to the vestibule. The membranous labyrinth within the cochlea contains the receptors for hearing.

Hearing and equilibrium receptors initiate nerve impulses which are carried by the vestibulocochlear nerve (cranial nerve VIII) to the brain. The **vestibulocochlear nerve** has two branches: the **vestibular branch** that carries nerve impulses generated by equilibrium receptors, and the **cochlear branch** that carries nerve impulses generated by the hearing receptors.

ACTIVITY 1 ANATOMY OF THE EAR

1 Label the structures of the external, middle, and internal ear in Figure 18.1.

2 Using contrasting colors, color the ear structures in Figure 18.1.

3 Identify the structures on a model or chart.

4 Pronounce the terms as you point to them.

5 Identify the area of the temporal bone that houses the middle and internal ear on a skull.

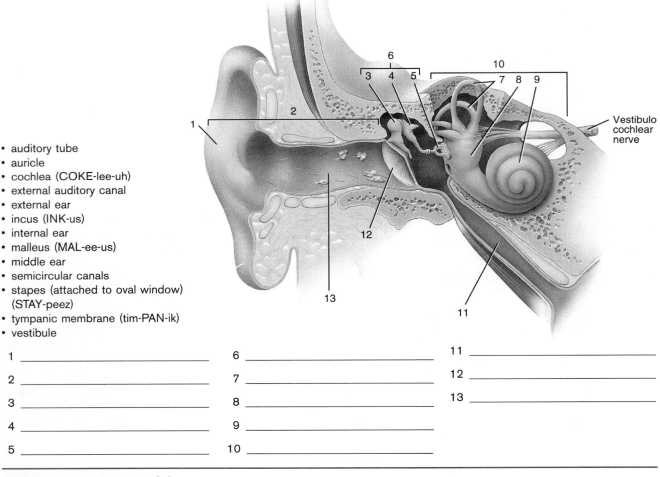

- auditory tube
- auricle
- cochlea (COKE-lee-uh)
- external auditory canal
- external ear
- incus (INK-us)
- internal ear
- malleus (MAL-ee-us)
- middle ear
- semicircular canals
- stapes (attached to oval window) (STAY-peez)
- tympanic membrane (tim-PAN-ik)
- vestibule

1 _____   6 _____   11 _____

2 _____   7 _____   12 _____

3 _____   8 _____   13 _____

4 _____   9 _____

5 _____   10 _____

FIGURE 18.1 **Anatomy of the ear.**

B. AUDITORY AND EQUILIBRIUM TESTS

Hearing loss can be described as either conduction deafness or sensorineural deafness. **Conduction deafness** occurs when there is a decreased ability to conduct the energy of sound waves through the external and middle ear to hearing receptors in the inner ear. Ear wax buildup, damage to the tympanic membrane, or fusion of auditory ossicles may cause conduction deafness. **Sensorineural deafness** is caused by damage to hearing receptors, damage to the cochlear branch of the vestibulocochlear nerve, or damage of the neural pathways to the auditory cortex.

Equilibrium receptors provide information that enables the body to maintain balance. There are two types of equilibrium receptors, static and dynamic. **Static equilibrium receptors** provide information about body position relative to the force of gravity (standing upright vs. being upside down). **Dynamic equilibrium receptors** provide information about body position in response to sudden movement such as rotation, acceleration, and deceleration (spinning, going faster, stopping). Inflammation of or injury to equilibrium receptors results in an inability to maintain body position, vertigo, and/or dizziness. **Vertigo** is the sensation of circular motion either of oneself or external objects, while dizziness is often used to describe faintness, unsteadiness, or lightheadedness. Severe vertigo may be accompanied by **nystagmus**—rapid, involuntary movement of eyeballs.

ACTIVITY 2 AUDITORY AND EQUILIBRIUM TESTS

1 Test for unilateral (one side only) deafness using the **Weber test.**

- Choose a subject from your group. Have the subject sit with the head erect and facing forward.
- Strike a tuning fork (middle C preferably) and place it medially on the subject's forehead (bone conduction).
- Ask the subject if the sound is equally loud in both ears or louder in one ear. Circle the result in Table 18.1.
- If the sound is equally loud in both ears, the subject either has normal hearing or bilateral deafness (equal hearing loss in both ears). If the subject hears the sound louder in one ear, then the subject may have unilateral deafness (either conduction deafness in that ear or sensorineural deafness in the opposite ear).
- If the subject has unilateral deafness, conduct the Rinne test to determine if hearing loss is conduction deafness.
- *Optional: Mimic unilateral conduction deafness by placing a cotton ball in one external auditory canal and repeat the Weber test.*

2 Test for conduction deafness using the **Rinne test.**

- This test will be conducted on the ear that may have conduction deafness as indicated by the Weber test.

TABLE 18.1 Results of Weber, Rinne, and Barnay Tests

TEST	RESULTS (CIRCLE YOUR RESULTS)		
Weber test	Equal loudness in both ears = normal hearing or equal hearing loss in both ears		
	or		
	Sound is louder in right ear = conduction deafness in right ear or sensorineural deafness in left ear		
	or		
	Sound is louder in left ear = conduction deafness in left ear or sensorineural deafness in right ear		
Rinne test	Conduction deafness *or* no conduction deafness		
Balance test	Static equilibrium receptors: functioning *or* not functioning		
		Direction of eye movement	Name semicircular canal stimulated
Barnay test	Head slightly forward:	lateral *or* vertical *or* rotational	lateral *or* anterior *or* posterior
	Head toward shoulder:	lateral *or* vertical *or* rotational	lateral *or* anterior *or* posterior
	Head on chin:	lateral *or* vertical *or* rotational	lateral *or* anterior *or* posterior

- Have the subject sit with head erect and facing forward.

- Strike a tuning fork and place it on the subject's mastoid process to test hearing by **bone conduction.**

- Ask the subject to tell you when the sound can no longer be heard. Immediately place the still-vibrating tuning fork close to the subject's ear to test hearing by air conduction. If the subject can hear the tuning fork again when it is placed next to his or her ear, the subject does not have conduction deafness in that ear. If the subject cannot hear the tuning fork again, then the subject *may* have conduction deafness in that ear.

- Circle the results in Table 18.1. If the subject does not have conduction deafness, the test is complete.

- To verify conduction deafness, test the same ear again. This time you are testing hearing by **air conduction** first.

- Strike the tuning fork again and place the tuning fork close to the subject's ear.

- Ask the subject to tell you when the sound can no longer be heard, then place the tuning fork on the subject's mastoid process (bone conduction). If the subject hears the sound again, there is conduction deafness in that ear. Record whether the subject has conduction deafness or no conduction deafness in that ear by circling the results in Table 18.1.

3 Conduct a balance test to evaluate **static equilibrium** receptors.

- Have the subject stand in front of a whiteboard or chalkboard with arms at the sides. The subject cannot lean against the wall or support him- or herself in any manner.

- Tell the subject to stand perfectly still. Mark the outline of the shoulders to help determine when he or she sways.

- Tell the subject to close his or her eyes. Observe movement in the subject's shoulders. Notice that, although the subject may sway slightly, posture is always corrected. Signals from the static equilibrium receptors are helping the subject to maintain posture. If the static equilibrium receptors are not functioning, the subject will not be able to maintain posture and will exhibit large swaying move-

ments or will fall. Be prepared to support the subject if necessary.

- Record the results in Table 18.1.

4 Conduct the **Barany test** to evaluate function of semicircular canals and **dynamic equilibrium** receptors.

- Choose a subject from your lab group who does not readily experience dizziness or become nauseated when rotated. If the subject experiences nausea during the demonstration, immediately stop rotation.

- Provide the subject with a chair or stool that can be rotated. Have the subject sit on the chair and hold onto the arms or seat for safety. Decide how the subject will position his or her legs during rotation to ensure safety and prevent interference. Position 3 to 4 students around the chair to prevent the subject from falling off the chair.

- Tell the subject to slightly tilt his or her head forward, focus on a distant object, and keep both eyes open during the rotation.

- Carefully turn the chair or stool clockwise (to the right). Complete 10 turns, one turn per 2 seconds and stop suddenly. **Be prepared to support the subject until vertigo and/or dizziness has passed.** The subject will still experience rotation, indicating that the semicircular canals are functioning. Endolymph continues to move within the membranous semicircular ducts for a short time after rotation has stopped.

- Observe which way the subject's eyeballs are moving immediately after stopping rotation. Record direction in Table 18.1.

    Lateral movement of the eyes indicates stimulation of the dynamic equilibrium receptors in the lateral semicircular canals.

    Vertical movement of the eyes indicates stimulation of the dynamic equilibrium receptors in the anterior semicircular canals.

    Rotational movement of the eyes indicates stimulation of the dynamic equilibrium receptors in the posterior semicircular canals.

- Repeat the demonstration with the subject's head tilted toward one shoulder, and then again with the subject's chin resting on his or her chest.

REVIEWING YOUR KNOWLEDGE

A. External and Middle Ear Structures

Write the name of the structure that applies to each statement. Terms may be used more than once.

auditory tube
auricle
external auditory canal
incus
malleus
oval window
stapes
tympanic membrane

_____ **1.** Auditory ossicle attached to tympanic membrane.

_____ **2.** Equalizes air pressure in middle ear with external air pressure.

_____ **3.** External ear structure that collects sound waves and directs them to external auditory canal.

_____ **4.** Ear drum.

_____ **5.** Conducts sound from auricle to tympanic membrane.

_____ **6.** Stapes is attached to this membrane-covered opening.

_____ **7.** Middle auditory ossicle.

_____ **8.**
_____ **9.** Small bones of middle ear that are connected by synovial joints.
_____ **10.**

B. Internal Ear Structures

Write the name of the structure that applies to each statement.

cochlea
endolymph
perilymph
semicircular canals
vestibule

_____ **1.** Fluid found within the membranous labyrinth.

_____ **2.** Fluid found within all bony labyrinth structures.

_____ **3.** Hearing receptors are found within the membranous labyrinth in this structure.

_____ **4.**

_____ **5.** } Equilibrium receptors are found within the membranous labyrinth in these structures.

C. Auditory and Equilibrium Tests

Answer the following questions.

1. What causes nystagmus?

2. What causes conduction deafness?

3. What causes sensorineural deafness?

4. If a vibrating tuning fork is placed on the mastoid process, who would "hear" the sound: someone with normal hearing, someone with conduction deafness, or someone with sensorineural deafness? Circle all that apply.

5. Inability to maintain posture while standing would indicate a problem with which equilibrium receptor?

ENDOCRINE SYSTEM

After completing this exercise, you should be able to:

- Identify each major endocrine gland on models or charts
- Name the principal hormones secreted by the major endocrine glands and describe their main function

- human torso with a dissectable brain or other models and charts depicting endocrine glands
- skull
- colored pencils

The **endocrine system** (*endo-* = within; *krinein* = to secrete) has many glands that secrete chemicals called hormones into the bloodstream. These chemicals are transported throughout the body in the blood and bind to **target cells.** Target cells for a hormone have specific **receptors** to which the hormone binds. Hormones cause changes in activity in these target cells to assist in maintaining homeostasis.

A. IDENTIFICATION OF MAJOR ENDOCRINE GLANDS

The major endocrine glands from the head inferiorly are the **pineal gland (body), hypothalamus, pituitary gland, thyroid gland, parathyroid glands, thymus, adrenal glands, pancreas, ovaries,** and **testes.**

1. Hypothalamus and Pituitary Gland

Located inferior to the thalamus in the brain, the **hypothalamus** is an important component of both the nervous and endocrine systems and couples these two regulatory systems together. Although in the past the pituitary gland was called the "master gland," it is now known that the pituitary gland is controlled by the hypothalamus. Hormones secreted by the hypothalamus either stimulate or inhibit secretion of anterior pituitary hormones.

The pituitary gland is attached to the inferior surface of the brain and is protected by the sella turcica of the skull. The **infundibulum** is the stalk that connects the hypothalamus to the pituitary gland. There are two lobes of the **pituitary gland,** the anterior pituitary (adenohypophysis) or anterior lobe and the posterior pituitary (neurohypophysis) or posterior lobe. The **anterior pituitary** is composed of glandular tissue that makes and secretes seven hormones: **human growth hormone** (hGH), **thyroid-stimulating hormone** (TSH), **follicle-stimulating hormone** (FSH), **luteinizing hormone** (LH), **prolactin** (PRL), **adrenocorticotropic hormone** (ACTH), and **melanocyte-stimulating hormone** (MSH). The **posterior pituitary** is smaller than the anterior pituitary and is composed of neural tissue that stores and releases two hormones into the bloodstream, **antidiuretic hormone** (ADH) and **oxytocin** (OT). These two hormones are synthesized by the hypothalamus, are packaged into vesicles, and travel down axons that pass through the infundibulum to the posterior pituitary gland.

2. Thyroid and Parathyroid Glands

The thyroid gland lies on both sides of the trachea near the larynx and has two lobes with a connecting **isthmus.** The thyroid gland synthesizes and secretes **calcitonin** and two **thyroid hormones (TH), thyroxine (T_4)** or **tetraiodothyronine** (*tetra-* = four; *iodo-* = iodine) and **triiodothyronine (T_3).** Embedded in the posterior surface of the thyroid gland are typically four small, round **parathyroid glands** that secrete **parathyroid hormone (PTH).**

3. Adrenal Glands

The **adrenal glands** (*ad-* = addition to; *renal* = kidneys) are also known as the suprarenal glands because of their location on the superior part of the kidneys. Each adrenal gland is surrounded by a **capsule** and is composed of a larger outer cortex and a central medulla. The cortex and medulla are comprised of different tissue types, and therefore produce and secrete distinct types of hormones. The **adrenal cortex** has three distinct zones with each secreting different types of steroid hormones: mineralcorticoids (**aldosterone**), glucocorticoids (**cortisol**), and **androgens.** The **adrenal medulla** contains nervous tissue that is stimulated by the sympathetic nervous system to secrete two hormones, **epinephrine and norepinephrine (NE).** Like the posterior pituitary gland, the chemicals secreted by this nervous tissue are called hormones because they are secreted into the bloodstream and travel to another part of the body to target cells.

4. Pancreas

The **pancreas** is composed of a head, body, and tail. Its head is cradled in the curvature of the duodenum (area of small intestine distal to the stomach). The body and tail of the pancreas are inferior and posterior to the stomach near the spleen. The pancreas has both endocrine and exocrine (*exo-* = outside; literally "to secrete outside") functions. The exocrine cells vastly outnumber the endocrine cells. Enzymes produced by exocrine cells are secreted into the duodenum of the small intestine through a duct called the **pancreatic duct** and function in digestion. Endocrine cells, called **pancreatic islets (islets of Langerhans),** are situated as little islands among the exocrine cells. **Alpha cells** in the pancreatic islets secrete **glucagon** and **beta cells** secrete **insulin.**

5. Ovaries, Testes, Pineal Gland, and Thymus

The **ovaries** or female gonads are located in the female pelvic cavity. They form ova and produce and secrete two major female steroid hormones, **estrogen** and **progesterone.** Located within the scrotum, the **testes** (male gonads) form sperm and produce and secrete androgens, primarily **testosterone.**

The **pineal gland** (*pinea-* = pinecone) or **pineal body** secretes the hormone **melatonin** and is a small, cone-shaped gland located in the brain posterior to the thalamus and superior to the cerebellum. The **thymus** secretes the hormone **thymosin** and is located anterior and superior to the heart. This gland is much larger in infants and children and regresses in size or atrophies as a person ages.

ACTIVITY 1 IDENTIFICATION OF THE MAJOR ENDOCRINE ORGANS

1 Using your text as necessary, label the major endocrine glands in Figure 19.1.

2 Using contrasting colors, color the endocrine organs in Figure 19.1.

3 Using Table 19.1, review the hormone(s) produced by each gland and the target cells and function of each hormone.

4 Label the hypothalamus and the major structures of the pituitary gland in Figure 19.2, the thyroid gland in Figure 19.3(a), the parathyroid glands in Figure 19.3(b), the adrenal gland in Figures 19.4(a) and (b), and the pancreas in Figure 19.5.

5 Identify the hypothalamus, pituitary gland, thyroid gland, parathyroid glands, adrenal glands, pancreas, ovaries, testes, pineal gland, and thymus on a model or chart.

6 Pronounce each of these terms as you point to them.

7 As you point to each gland, name each hormone it secretes, the target cells of each hormone, and the function of each hormone.

8 Examine the interior of a skull and locate the sella turcica that protects the pituitary gland.

9 Palpate the thyroid gland on yourself. Place your fingers on each side of the trachea, three fingers' breadth superior to the sternum, and feel the thyroid gland move upward as you swallow.

- adrenal glands (a-DREE-nul)
- hypothalamus (hypo-THAL-a-mus)
- ovaries
- pancreas
- parathyroid glands (para-THY-roid)
- pineal gland (pie-NEE-ul)
- pituitary gland or hypophysis (hy-POF-ih-sis)
- testes (TES-teez)
- thymus (THY-mus)
- thyroid gland

1 _____

2 _____

3 _____

4 _____

5 _____

6 _____

7 _____

8 _____

9 _____

10 _____

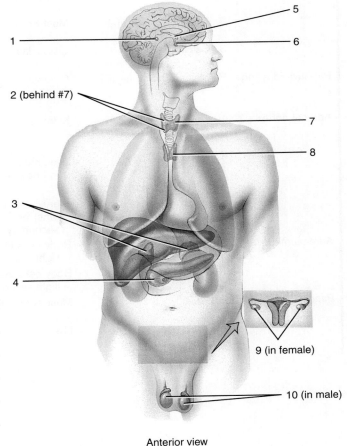

Anterior view

FIGURE 19.1 **Major endocrine glands.**

TABLE 19.1 Endocrine Glands, Hormones, Target Cells, and Hormone Function

GLAND	HORMONE	LOCATION OF TARGET CELLS	HORMONE FUNCTION
Anterior pituitary	1. Human growth hormone (hGH)	Cartilage, bone, skeletal muscle, liver, and other body tissues	Stimulates secretion of hormones that stimulate body growth and metabolism.
	2. Thyroid-stimulating hormone (TSH)	Thyroid gland	Stimulates growth of thyroid gland and secretion of its hormones.
	3. Follicle-stimulating hormone (FSH)	Testes	Stimulates sperm production.
		Ovaries	Stimulates oocyte production and estrogen secretion.
	4. Luteinizing hormone (LH)	Testes	Stimulates secretion of testosterone.
		Ovaries	Triggers ovulation and stimulates secretion of estrogen and progesterone.
	5. Prolactin (PRL)	Mammary gland	Stimulates production and secretion of milk.
	6. Adrenocorticotropic hormone (ACTH)	Adrenal cortex	Stimulates secretion of hormones by adrenal cortex.
	7. Melanocyte-stimulating hormone (MSH)	Skin	Darkens skin pigmentation.
Posterior pituitary	1. Antidiuretic hormone (ADH)	Kidneys	Decreases water lost in urine by returning water to the blood.
	2. Oxytocin (OT)	Uterus and mammary glands	Stimulates uterine contractions and milk ejection during suckling.
Thyroid gland	1. Thyroxine (T_4)	Most body cells	Increases metabolism and basal metabolic rate (BMR).
	2. Triiodothyronine (T_3)	Most body cells	Increases metabolism and BMR.
	3. Calcitonin	Osteoclast cells in bones	Decreases blood calcium levels by inhibiting osteoclasts.
Parathyroid glands	Parathyroid hormone (PTH)	Osteoclast cells in bones	Increases blood calcium levels by stimulating osteoclasts to break down bone matrix.
Adrenal cortex	1. Aldosterone	Kidneys	Decreases sodium and water loss in urine by returning sodium and water to the blood.
	2. Cortisol	Liver, muscle, and cells involved in body defenses	Increases resistance to stress, increases blood glucose levels, and decreases inflammation.
	3. Androgens	Uterus, mammary glands, and other body cells involved in secondary sex characteristics	Insignificant in males; increases sex drive in females.
Adrenal medulla	1. Epinephrine	Body cells involved in fight-or-flight response	Promotes fight-or-flight response.
	2. Norepinephrine (NE)	Body cells involved in fight-or-flight response	Promotes fight-or-flight response.
Pancreas	1. Insulin	Most body cells	Decreases blood glucose levels by transporting glucose into body cells.
	2. Glucagon	Liver	Increases blood glucose levels by stimulating liver to break down glycogen into glucose.

TABLE 19.1 Endocrine Glands, Hormones, Target Cells,...(*continued*)

GLAND	HORMONE	LOCATION OF TARGET CELLS	HORMONE FUNCTION
Ovaries	1. Estrogen	Uterus, mammary glands, and other body cells involved in female sexual characteristics	Stimulates development of female sex characteristics; helps regulate menstrual cycle.
	2. Progesterone	Uterus, mammary glands and other body cells involved in female sexual characteristics	Stimulates development of female sex characteristics; helps regulate menstrual cycle.
Testes	Testosterone	Testes, muscle, and other body cells involved in male sexual characteristics	Stimulates development of male sex characteristics; stimulates male sex drive; regulates sperm production.
Pineal Gland	Melatonin	Brain	Helps to set biological clock.
Thymus	Thymosin	T cells (type of white blood cell involved in immune response)	Promotes the maturation of T cells for the immune response.

- anterior pituitary or adenohypophysis (a-den-oh-hy-POF-ih-sis)
- hypothalamus (hypo-THAL-uh-mus)
- infundibulum (in-fun-DIB-u-lum)
- posterior pituitary or neurohypophysis (neur-oh-hy-POF-ih-sis)

1 _____

2 _____

3 _____

4 _____

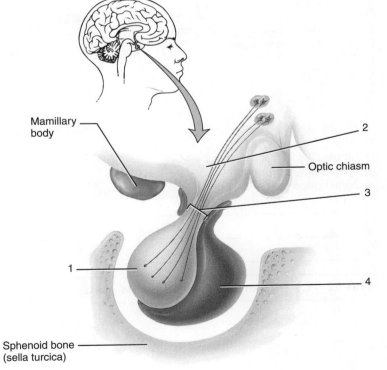

FIGURE 19.2 **Hypothalamus and pituitary gland.**

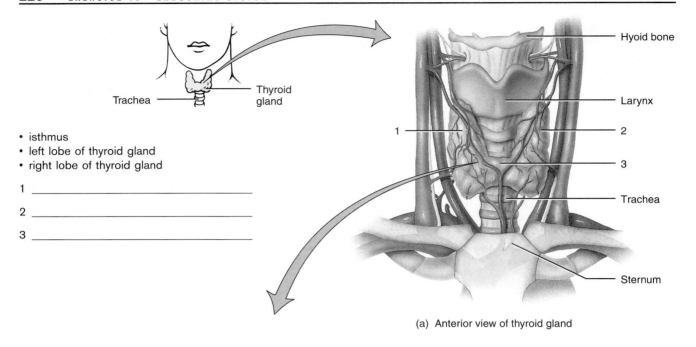

- isthmus
- left lobe of thyroid gland
- right lobe of thyroid gland

1 _____

2 _____

3 _____

(a) Anterior view of thyroid gland

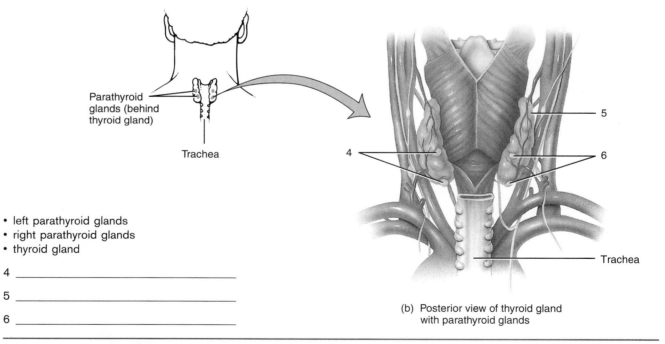

- left parathyroid glands
- right parathyroid glands
- thyroid gland

4 _____

5 _____

6 _____

(b) Posterior view of thyroid gland
     with parathyroid glands

FIGURE 19.3   The thyroid and parathyroid glands.

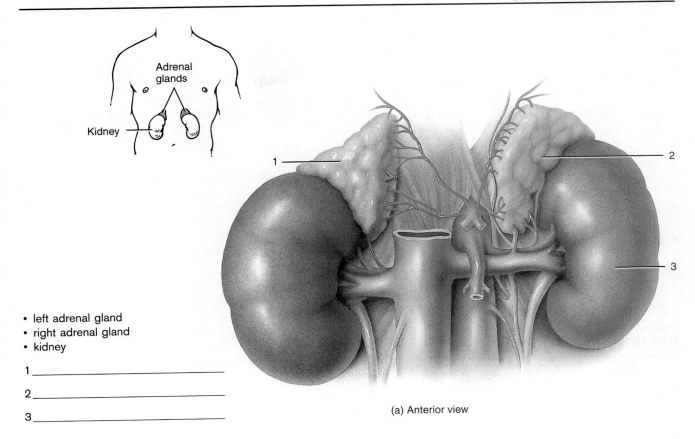

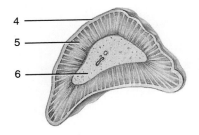

(a) Anterior view

- left adrenal gland
- right adrenal gland
- kidney

1 _____

2 _____

3 _____

(b) Section through left adrenal gland

- adrenal cortex
- adrenal medulla
- capsule

4 _____

5 _____

6 _____

FIGURE 19.4   The adrenal glands.

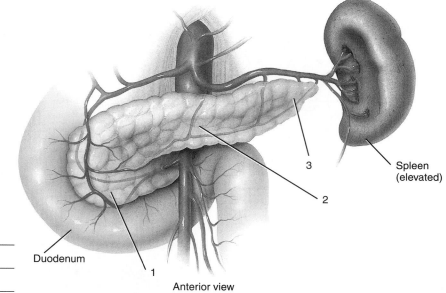

- body of pancreas
- head of pancreas
- tail of pancreas

1 _____

2 _____

3 _____

Duodenum

Anterior view

Spleen (elevated)

3

2

1

FIGURE 19.5 **The pancreas.**

REVIEWING YOUR KNOWLEDGE

A. Hormone Abbreviations

Write the name of each hormone next to its abbreviation.

1. ACTH _____

2. ADH _____

3. FSH _____

4. hGH _____

5. LH _____

6. OT _____

7. PRL _____

8. PTH _____

9. T_3 _____

10. T_4 _____

11. TSH _____

12. TH _____

B. Main Endocrine Organs and Their Hormones

Write the name of the endocrine gland that *secretes* the following hormones.

adrenal cortex	pineal gland
adrenal medulla	posterior pituitary
anterior pituitary	testes
ovaries	thyroid gland
pancreas	thymus
parathyroid glands	

Hormone	**Endocrine Gland**
1. ACTH	_____
2. ADH	_____
3. aldosterone	_____
4. androgens	_____
5. calcitonin	_____
6. cortisol	_____
7. epinephrine/NE	_____
8. estrogen; progesterone	_____
9. FSH	_____
10. glucagon	_____
11. hGH	_____
12. insulin	_____
13. LH	_____
14. melatonin	_____
15. OT	_____
16. PRL	_____
17. PTH	_____
18. T_3 and T_4	_____
19. testosterone	_____
20. TSH	_____
21. thymosin	_____
22. MSH	_____

C. Hormone Function

Write the name of the hormone that matches its function.

ACTH	glucagon	progesterone
ADH	hGH	PTH
aldosterone	insulin	T_3
androgens	LH	T_4
calcitonin	melatonin	testosterone
cortisol	MSH	thymosin
epinephrine	NE	TSH
estrogen	OT	
FSH	PRL	

Hormone Function

_____ **1.** Stimulates uterine contractions and milk ejection during suckling.

_____ **2.** Stimulates secretion of hormones by the adrenal cortex.

_____ **3.** Decreases water loss by increasing reabsorption of water into blood and decreasing urine production.

_____ **4.** Increases sex drive in females.

_____ **5.**
                         } Increases metabolism and BMR.
_____ **6.**

_____ **7.** Triggers ovulation and stimulates secretion of estrogen and progesterone.

_____ **8.** Increases blood calcium levels by stimulating osteoclast activity.

_____ **9.**
                         } Promotes fight-or-flight response.
_____ **10.**

_____ **11.** Stimulates production and secretion of milk.

_____ **12.** Increases resistance to stress, increases blood glucose levels, and decreases inflammation.

_____ **13.** Helps to set the biological clock.

_____ **14.** Darkens skin pigmentation.

_____ **15.** Stimulates secretion of hormones that stimulate body growth and metabolism.

_____ **16.** Decreases blood calcium levels by inhibiting osteoclasts.

_____ **17.** Promotes the maturation of T cells for the immune response.

_____ **18.** Stimulates oocyte production and estrogen secretion.

_____ **19.** Decreases blood glucose levels by transporting glucose into body cells.

_____ **20.** Increases blood glucose levels by stimulating the liver to break down glycogen into glucose.

_____ **21.**

Stimulates development of female sex characteristics and helps regulate menstrual cycle.

_____ **22.**

_____ **23.** Regulates sperm development, stimulates development of male sex characteristics, and stimulates male sex drive.

_____ **24.** Stimulates secretion of testosterone.

_____ **25.** Stimulates secretion of thyroid hormones.

_____ **26.** Stimulates sperm production.

_____ **27.** Increases reabsorption of sodium and water into blood and decreases urine output.

THE BLOOD

20

After completing this exercise, you should be able to:

- Name the components of blood
- Describe the structure, characteristics, and function of red blood cells (RBCs), white blood cell (WBCs), and platelets
- Identify RBCs, WBCs, and platelets on prepared blood smear slides
- Describe the importance of and perform ABO and Rh blood typing
- Describe the importance of the following blood tests: differential WBC count, hematocrit, hemoglobin, and coagulation time.

- prepared slides of normal human blood (with Wright's stain)
- compound microscopes and lens paper
- blood typing kit or antisera (anti-A; anti-B; anti-D), new test cards or glass slides (2 per student), toothpicks, wax marking pencil, Rh warming tray
- materials needed if using students' blood or purchased blood: biohazardous waste container; sharps container; safety glasses; disposable gloves; 10% bleach solution in spray bottle; sterile, disposable lancets or Autolets; alcohol swabs; cotton balls

A. COMPONENTS OF BLOOD

The adult cardiovascular system contains approximately 5.5 liters of blood. When centrifuged, **blood** separates visually into two main components, plasma and formed elements. The clear, straw-colored liquid is called **plasma** and the dark red and buff-colored portions are the **formed elements** (Figure 20.1). The formed elements include **red blood cells** (RBCs), **white blood cells** (WBCs), and **platelets** (cell fragments). Red blood cells are also called **erythrocytes** (*erythro-* = red; *-cytes* = cells), WBCs are known as **leukocytes** (*leuko-* = white), and platelets are known as **thrombocytes** (*thrombo-* = clot). The formed elements constitute approximately 45% of whole blood volume, and plasma composes about 55%. Plasma is about 91.5% water and 8.5% solutes. The solutes are mostly plasma proteins, but also include nutrients (glucose, amino acids, and lipids), blood gases (oxygen and carbon dioxide), electrolytes, hormones, enzymes, and waste materials.

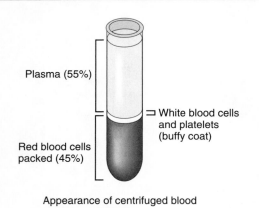

Plasma (55%)

White blood cells and platelets (buffy coat)

Red blood cells packed (45%)

Appearance of centrifuged blood

FIGURE 20.1   Components of blood in a normal adult.

B. THE STRUCTURE AND FUNCTION OF RED BLOOD CELLS, WHITE BLOOD CELLS, AND PLATELETS

1. Red Blood Cells (Erythrocytes or RBCs)

Erythrocytes are small, anucleate (without a nucleus) cells that contain **hemoglobin,** a large molecule used to transport oxygen and carbon dioxide in the blood. About 33% of the total weight of RBCs is composed of hemoglobin. Hemoglobin contains a red pigment called heme that gives blood its red color. Blood is bright red when oxygen-rich and darker red when oxygen-poor. There are approximately 4.5 to 5 million RBCs per microliter (μL) of blood, with lower amounts in females and higher amounts in males. Having an abnormally high number of RBCs is called polycythemia (*poly-* = many; *cyto-* = cells; *-emia* = blood), and having an unusually low number of RBCs is one type of **anemia.**

2. White Blood Cells (Leukocytes or WBCs)

White blood cells are nucleated cells that attack pathogens and other substances that invade the body. There are five kinds of leukocytes that are divided into two categories, granular and agranular. **Granular leukocytes** have discernible vesicles (granules) in the cytoplasm that can be seen after staining. **Agranular leukocytes** also have granules, but the microscopes that were used at the time these cells were named were not powerful enough to distinguish the granules. Granular leukocytes include **neutrophils, eosinophils,** and **basophils.** Agranular leukocytes include **lymphocytes** and **monocytes.** There are normally 5,000 to 10,000 WBCs per microliter of blood. An abnormally high number of WBCs is called **leukocytosis** (*-osis* = an increase in a pathological condition), and an abnormally low number of WBCs is called **leukopenia** (*-penia* = deficiency).

3. Platelets (Thrombocytes; *thrombo-* = clot)

Platelets are formed from large, multinuclear cells called **megakaryocytes** (*mega-* = large; *karyo-* = nucleus). Megakaryocytes break into small, disc-shaped fragments called platelets that do not have nuclei, and are therefore not considered to be cells. These special cell fragments protect the body by forming a platelet plug to stop bleeding when blood vessels rupture and by secreting chemicals that aid in blood clotting. **Thrombocytopenia** is a deficiency in the number of circulating platelets.

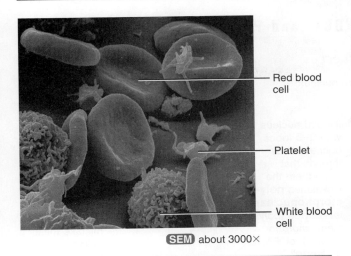

Red blood cell

Platelet

White blood cell

SEM about 3000×

FIGURE 20.2   **Scanning electron micrograph of the formed elements of blood.**

4. Microscopic Examination of the Formed Elements

Blood cells are difficult to see at low magnifications because the cells are so small. The erythrocytes are the most numerous and are small, pinkish cells. The WBCs are not very numerous but are conspicuous because their nuclei stain dark blue to dark purple with a Wright's stain.

Because the majority of WBCs are larger than the RBCs, you can use RBCs as a standard to compare the size of WBCs. The tiny platelets stain dark purple, are dispersed among the RBCs and WBCs, and may look to you like extra stain on the slide. Keep in mind that blood cells have a three-dimensional spherical shape. You will not always find the nucleus in WBCs appearing as described or as seen in photomicrographs because of the various views that are possible. Figure 20.2 depicts the three-dimensional structure of RBCs, WBCs, and platelets.

ACTIVITY 1 IDENTIFICATION OF RBCs, WBCs, AND PLATELETS

1 Examine a prepared blood slide to locate RBCs and WBCs.

 • Use the low-power lens to locate RBCs and WBCs.

 • Use the high-power or oil-immersion lens to locate the platelets and to identify the different types of WBCs.

2 To identify the WBC type, refer to Table 20.1 and Figure 20.3. Look at the size of the cell compared with RBCs, the presence and color of granules, the shape and number of lobes to the nucleus, and the color of the cytoplasm.

TABLE 20.1 Characteristics of RBCs, WBCs, and Platelets (Wright Stain)

	RELATIVE NUMBER	SIZE	NUCLEUS	GRANULES	COLOR OF CYTOPLASM
RBCs	700:1 ratio of RBCs to WBCs	7–8 micrometer diameter	No nucleus	No granules	Light pinkish-red color
Neutrophils (*neutr-* = neutral; *-phil* = loving)	60–70% of all WBCs	About 10–12 micrometer diameter	Multilobed nucleus with 2–5 lobes connected by threads that gives them the nicknames **polymorphonuclear** leukocytes (*poly-* = many; *morpho-* = shapes), or **PMNs** or **"segs"** (segmented)	Small, nondistinct pale lilac to neutral-staining granules	Cytoplasm is usually pink but, depending on the stain used, sometimes has a reddish tinge that makes students mistake these cells for eosinophils
Eosinophils (*eosin-* = red, acidic dye)	2–4% of all WBCs	About 10–12 micrometer diameter	Bi-lobed nucleus (occasionally three lobes)	Many medium, red-orange granules (some stains make them look very dark brownish-red-orange)	Cytoplasm pink
Basophil (*baso-* = dark blue, basic dye)	0.5–1% of all WBCs	8–10 micrometer diameter	Nucleus large, varied in shape; generally obscured by large granules	Large, dark blue-purple granules	Cytoplasm appears purple
Lymphocyte (lymph cell)	20–25% of all WBCs	6–9 micrometer diameter (small lymphocyte); 10–14 micrometer diameter (large lymphocyte)	Large, round, or slightly indented nucleus; stains very dark purple	Granules not obvious with light microscope	Light sky-blue cytoplasm; small cells have only a rim of cytoplasm; more cytoplasm in larger lymphocytes
Monocyte (*mono-* = one; pertains to having only one nucleus)	3–8% of all WBCs	Very large cell; 12–20 micrometer diameter	Large kidney bean or horseshoe-shaped lacy nucleus; sometimes oval and indented	Granules not obvious with light microscope	Light blue-gray cytoplasm; sometimes irregular extensions of the cytoplasm
Platelets	20:1 ratio of RBCs to platelets	2–4 micrometer diameter; small cell fragments	No nucleus	Dark purple granules	Difficult to see because of dark purple granules

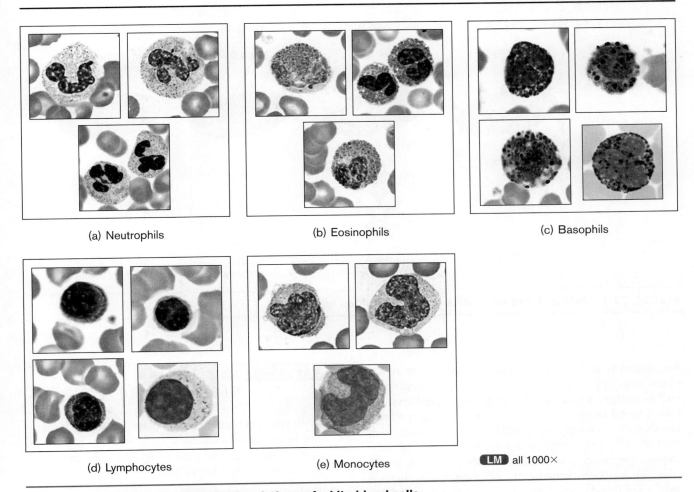

(a) Neutrophils

(b) Eosinophils

(c) Basophils

(d) Lymphocytes

(e) Monocytes

LM all 1000×

FIGURE 20.3 **Photomicrographs of variations of white blood cells.**

C. ABO AND Rh BLOOD TYPING

Blood typing is critical for blood transfusions, transplantations, and maternal–fetal compatibility, but it is also used in genetic studies, forensic studies, legal medicine, and anthropology. Although there are many different systems for classifying human blood, we will be studying the **ABO** and **Rh systems** because they are most commonly used. Blood typing is based on the **antigenic** molecules (agglutinogens) that are on the surface of the RBC membranes. An **antigen** is a substance that will react to a specific antibody to produce an immune response. In the **ABO system,** there are two types of antigens (A and B) that can be present as surface membrane molecules on RBCs. If the plasma membrane of your RBCs have only the A antigen present, you have **type A** blood; correspondingly, if you have only B antigens present, you have **type B** blood. If you have both antigens A and B present, you have **type AB** blood, and if you do not have either

A or B antigen present, you have **type O** blood (see Figure 20.4).

Antibodies are plasma proteins that combine with a specific antigen to inhibit or destroy it. ABO antibodies appear in babies' blood a few months after birth. If you have type A blood, you do not have its corresponding anti-A antibody. People with type A blood have anti-B antibodies that will become cross-linked and **agglutinate** (clump) if type B blood (contains B antigens) is given to them (Figure 20.5). Agglutination is followed by the activation of another plasma protein that attaches to the recipient's RBCs and hemolyzes or bursts them, releasing hemoglobin that can cause kidney damage. ABO antibodies do not cross the placenta because of their large size.

The **Rh blood system** is different from the ABO system but has some similarities. If you have the Rh antigen as a surface membrane molecule on your RBCs, you are **Rh-positive (Rh+).** If you do not have the Rh antigen, you are **Rh-negative (Rh−).** Those who are (Rh−) are not born with the anti-Rh antibody (as in the ABO system) and do not obtain this antibody until they are exposed to the

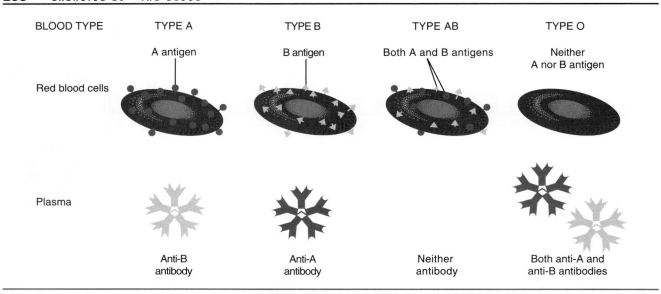

FIGURE 20.4 **Antigens and antibodies of the ABO blood types.**

Rh antigen from Rh+ blood. This can happen through a blood transfusion, by sharing hypodermic needles, or by an Rh- mother carrying an Rh+ child. During delivery, the baby's blood can leak from the placenta into the mother's bloodstream, causing the mother's body to make Rh antibodies. The first baby would not be affected, but subsequent pregnancies with Rh+ fetuses can result in the small Rh antibodies crossing the placenta causing hemolysis in the blood of the fetus. This condition is called **hemolytic disease of the newborn.** Rh- mothers are typically given a drug called RhoGAM so they will not make Rh antibodies.

The following activity will use **antisera** (plural of antiserum), which can be artificial serum or serum from an animal or human containing antibodies against A, B, or D (Rh) antigens. **Serum** is blood plasma without clotting proteins. If anti-A serum clumps a particular blood sample, the blood is type A because an antibody–antigen complex formed between the A antigens on the RBCs and the A antibodies in the antiserum.

ACTIVITY 2 ABO AND Rh BLOOD TYPING

1 With your lab partner(s), complete Table 20.2 by writing in the antibody present in plasma for each blood type. Also, determine each compatible and incompatible donor for each blood type.

2 Follow the safety precautions outlined here to familiarize yourself with blood handling safety procedures.

- When using human or animal blood samples, you must protect yourself from any blood-borne infectious disease (hepatitis, HIV, etc.).

- Wear gloves and safety goggles while performing blood tests.

- Place blood slides and toothpicks in a biohazard container according to your instructor's directions.

- Wash tabletops thoroughly with 10% bleach solution at the end of the lab.

TABLE 20.2 Summary of ABO Blood Group Interactions

BLOOD TYPE	A	B	AB	O
Antigen on RBCs				
Antibody in plasma				
Compatible donor blood types (no hemolysis)				
Incompatible donor blood types (hemolysis)				

- Wash your hands with soap and water before leaving the lab.

- If a blood spill occurs, cover the spill with paper towels soaked in 10% bleach solution and immediately inform your instructor.

3 Using Figure 20.5 as a guide, perform ABO blood typing according to your kit instructions or the following procedure.

- Obtain a clean glass slide (test cards if using a typing kit), two new toothpicks, a wax marking pencil, and anti-A and anti-B sera.

- Your instructor will tell you whether you will be using sterile blood, your own blood, or simulated blood.

- Divide the glass slide in half with a wax pencil. Mark A on the left side and B on the right side. Place one drop of anti-A serum on the left side and one drop of anti-B serum on the right side.

- Place one drop of blood on sides A and B of slide. If using your own blood, follow the procedure in #5 of this activity.

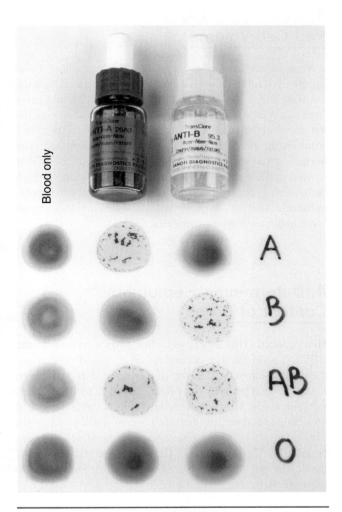

FIGURE 20.5 Serum agglutination results.

- Using *separate* new toothpicks, mix each sample of blood with the corresponding antiserum.

- Results may take up to 2 minutes. Observe each sample for agglutination or granulation (Figure 20.5) and interpret the results.

- Interpret the results and record them in Table 20.3.

- Write your results on the board to pool class data for comparison.

- Record percentage of ABO blood types for your class in Table 20.3.

4 Perform Rh blood typing according to the following procedure.

- Obtain a clean glass slide and anti-D serum (Rh).

- Place one drop of anti-D on the glass slide and add a drop of blood. If using your own blood, follow the procedure in #5 of this activity.

- Mix the two liquids with a new toothpick.

- Place slide on the warm Rh typing box and rock gently to mix. Rh typing requires a higher temperature than ABO typing does.

- Results may take up to 2 minutes. Observe your sample for clumping or granulation (Figure 20.5) and interpret the results.

- Sometimes it is more difficult to obtain positive results with the anti-Rh serum, depending on the supply and delivery situations.

- Dispose of blood slides and materials according to your instructor's directions.

- Record your results in Table 20.3 for each sample you tested.

- Write your results on the board to pool class data for comparison.

- Record percentage of Rh blood types for your class in Table 20.3.

5 Procedure for obtaining blood sample.

- Thoroughly wash hands with soap and dry with a clean paper towel.

- Thoroughly clean the tip of your index finger with an alcohol swab.

- Open a new, sterile blood lancet exposing the sharp tip only (or use an Autolet). Lance just to the side of the fingerpad with the new lancet. **Never** reuse a lancet, even your own.

- Deposit the used lancet in the sharps container for biohazard materials only.

- Wipe away the first drop of blood with a cotton ball and dispose in a biohazard waste container. Gently squeeze one drop of blood on each side of a clean prepared slide.

TABLE 20.3 Blood Typing Results		
SERUM AGGLUTINATION RESULTS	**YES**	**NO**
Clumping with anti-A		
Clumping with anti-B		
Clumping with anti-D (Rh)		
ABO blood type _____		
Rh type _____		
BLOOD TYPES	**CLASS BLOOD TYPES (PERCENTAGE)**	
A		
B		
AB		
O		
RH+		
RH−		

D. BLOOD TESTS

1. Complete Blood Count

A **complete blood count (CBC)** is a vital diagnostic tool that screens for abnormalities in the number or structure of blood cells. The CBC includes a total RBC count, total WBC count, platelet count, differential WBC count, hematocrit, and hemoglobin concentration. The CBC is used along with a battery of blood chemistry tests to put together a comprehensive profile of a person's general level of health based on blood values. Consult Exhibit 20.1 to observe the complexity of a CBC.

2. Differential White Blood Cell Count

A **differential WBC count** is performed to determine the percentage of each of the five types of WBCs in a blood sample. When attempting to learn which WBCs are normally present from the greatest to least percentages, it is helpful to remember the following mnemonic: "**N**ever **L**et **M**onkeys **E**at **B**ananas" (**N** = neutrophils; **L** = lymphocytes; **M** = monocytes; **E** = eosinophils; and **B** = basophils). Because the normal percentage of each WBC type is known, any significant abnormality in these percentages, elevated or depressed, can be indicative of particular disorders. Numerical values for the normal range of WBCs may vary depending on the reference source.

3. Hematocrit (Packed Red Cell Volume)

A **hematocrit** is the percentage of RBCs in a whole blood sample. A capillary tube of blood is centrifuged to pack the red cells at the outer end of the tube, separating them from the WBCs, platelets, and plasma. After measuring the length of the RBC column and the total length of the blood column, the percentage of RBCs can be calculated. The normal hematocrit range for females is approximately 38% to 46% and for males is 40% to 54%. An abnormally high hematocrit (generally 65% or above) is indicative of **polycythemia,** and a hematocrit with RBCs below the normal level indicates a type of **anemia.**

4. Determining Hemoglobin Content of Blood

Hemoglobin (Hb) is a protein that carries oxygen in the RBCs. Therefore, Hb concentration in blood determines the oxygen-carrying capacity of the blood. It is possible to have a normal hematocrit or RBC volume and still be anemic due to low Hb concentration. Normal values are 12 to 15g Hb per 100 mL of blood in females and 13 to 16g Hb per 100 mL in males. In severe anemia, the Hb content can be less than 7g per 100 mL. Clinically, the relationship of the hematocrit (%) is compared with the Hb reading (g/100 mL) and is generally a ratio of 3:1.

EXHIBIT 20.1 Complete Blood Count Form

PATIENT:		DOCTOR:
ACCOUNT NO:		REPORT FOR LOCATION:
MEDICAL RECORD NO.:		

HEMATOLOGY

DAY:	−1			REFERENCE	UNITS
DATE:	AUG 20 01				
TIME:	1000				
WBC	8.0			4.5–11.0	$\times 10^3$/UL
RBC	4.12	L		4.70–6.10	$\times 10^6$/UL
HGB	14.4			14.0–18.0	G/DL
HCT	41.6	L		42.0–52.0	%
MCV	101	H		83–99	FL
MCH	34.8	H		28–34	PG
MCHC	34.5			31.0–37.0	%
RDW	13.1			11.0–15.5	%
PLT	265			150–450	$\times 10^3$/UL
MPV	8.2			7.4–10.4	FL
NEUT%	47.7	L		49.0–90.0	%
LYMPH%	41.0			22.0–51.1	%
MONO%	6.5			0.0–13.0	%
EO%	4.3			0.0–7.0	%
BASO%	0.5			0.0–3.0	%
NEUT #	3.90			1.50–7.06	$\times 10^3$/UL
LYMPH #	3.30	H		0.70–3.00	$\times 10^3$/UL
MONO #	0.5			0.0–1.4	$\times 10^3$/UL
EO #	0.3			0.0–0.8	$\times 10^3$/UL
BASO #	0.0			0.0–0.3	$\times 10^3$/UL

COAGULATION

DAY:	−1		REFERENCE	UNITS
DATE:	AUG 20 01			
TIME:	1000			
PROTIME:	11.4		10.9–13.3	sec
INR:	1.0(a)			
PTT:	28.3		25.9–35.3	sec

NOTES: (a) INR pertains only to patient on anticoagulant therapy. INR is not relevant for patients not on anticoagulant therapy, nor is the stated therapeutic range applicable.

WBC = white blood cell; RBC = red blood cell; HGB = hemoglobin; HCT = hematocrit; MCV = mean corpuscular volume; MCH = mean corpuscular hemoglobin; MCHC = mean corpuscular hemoglobin concentration; RDW = red blood cell distribution index; PLT = platelet; MPV = mean platelet volume; NEUT% = neutrophil percent; LYMPH% = lymphocyte percent; MONO% = monocyte percent; EO% = eosinophil percent; BASO% = basophil percent; NEUT# = neutrophil number; LYMPH# = lymphocyte number; MONO# = monocyte number; EO# = eosinophil number; BASO# = basophil number; L = low; H = high; PTT = partial thromboplastin time.

5. Coagulation Time

The process of blood clotting or **coagulation** prevents excessive blood loss. There are many blood clotting factors present in the plasma. Other factors are released by injured tissues and platelets to initiate the chemical chain reaction that forms a blood clot. Fibrin is a long, insoluble thread-like protein strand that forms a mesh to trap platelets. Fibrin is formed from soluble fibrinogen molecules during clotting.

Coagulation time is the time it takes for clotting to take place when blood is removed from the body. Normal clotting time is within 2 to 6 minutes.

ACTIVITY 3 DIFFERENTIAL WBC COUNT, HEMATOCRIT, HEMOGLOBIN CONTENT OF BLOOD, AND COAGULATION TIME

1 Using the Complete Blood Count Form (Exhibit 20.1), find the percentage of WBC's, hematocrit percentage (HCT), and hemoglobin content of blood (HGB). Record these values in Table 20.4. Note: This patient is a male.

2 Calculate the hematocrit for blood in the capillary tubes in Figure 20.6.

- Use a millimeter ruler to measure (in mm) the length of the whole column (the length of packed RBCs, WBCs, and plasma) and record in Table 20.5.

- Measure the length (in mm) of the RBCs and record in Table 20.5.

- Use the following formula to calculate the percentage of RBCs (the hematocrit) in whole blood.

$$\frac{\text{Length of RBCs in mm}}{\text{Length of the whole column in mm}} \times 100$$

- Record the hematocrit in Table 20.5.

3 Answer discussion questions with your group.

TABLE 20.4 Blood Test Results

A. DIFFERENTIAL WBC COUNT

TYPE OF WBC	WBC (%)	NORMAL PERCENTAGE FOR WBCs
Neutrophil		60–70%
Lymphocyte		20–25%
Monocyte		3–8%
Eosinophil		2–4%
Basophil		0.5–1%

B. HEMATOCRIT AND HEMOGLOBIN CONTENT

HEMATOCRIT (%)	NORMAL RANGE	HEMOGLOBIN CONTENT	NORMAL RANGE

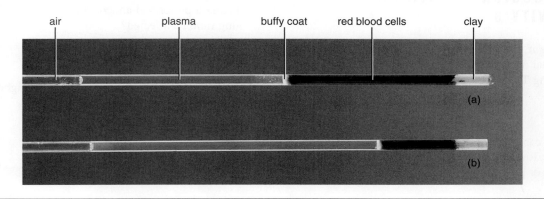

air   plasma   buffy coat   red blood cells   clay

(a)

(b)

FIGURE 20.6   Capillary tubes for calculating a hematocrit.

TABLE 20.5 Hematocrit Results

HEMATOCRIT CAPILLARY TUBES	RBC COLUMN (mm)	WHOLE COLUMN (mm)	HEMATOCRIT (PERCENTAGE OF RBCS/TOTAL BLOOD)
Hematocrit a			
Hematocrit b			
Hematocrit (your blood)			

MALE CLASS AVERAGE	FEMALE CLASS AVERAGE	NATIONAL AVERAGE FOR MALES	NATIONAL AVERAGE FOR FEMALES
		40–50%	38–46%

TABLE 20.6 Significance of Elevated and Depressed WBCs

WBC TYPE	HIGH COUNT MAY INDICATE	LOW COUNT MAY INDICATE
Neutrophils	Bacterial infection, burns, stress, inflammation	Radiation exposure, drug toxicity, vitamin B_{12} deficiency, systemic lupus erythematosus
Lymphocytes	Viral infections, some leukemias	Prolonged illness, immunosuppression, treatment with cortisol
Monocytes	Viral or fungal infections, tuberculosis, some leukemias, other chronic diseases	Bone marrow depression, treatment with cortisol
Eosinophils	Allergic reactions, parasitic infections, autoimmune diseases	Drug toxicity, stress
Basophils	Allergic reactions, leukemias, cancers, hypothyroidism	Pregnancy, ovulation, stress, hyperthyroidism

❓ DISCUSSION QUESTIONS FOR ACTIVITY 3

1 Indicate if the differential WBC count in Exhibit 20.1 is normal or abnormal. If abnormal, indicate the disorder using Table 20.6.

2 For the capillary tubes in Figure 20.6, is the hematocrit normal, low, or high? If abnormal, do values indicate anemia or polycythemia?

3 Compare the hematocrit results in Exhibit 20.1 with the grams of Hg per 100 mL. Calculate the ratio. _____ Is this close to the normal ratio? _____

4 Anticoagulants inhibit blood clotting. If a person was taking a prescribed anticoagulant, how would the clotting time be affected?

5 Would a person with hemophilia have higher or lower than normal clotting times?

EXERCISE

20

REVIEWING YOUR KNOWLEDGE

A. Characteristics of Formed Elements

Fill in the blank with the correct answer.

erythrocytes
leukocytes
megakaryocytes
thrombocytes

_____ **1.** The oxygen and carbon dioxide carrying cells.

_____ **2.** Help the body fight infections and foreign substances.

_____ **3.** Form a clot to help the body stop bleeding.

_____ **4.** Large cells that develop into platelets.

B. Blood Abnormalities

Fill in the blank with the correct answer.

anemia
leukocytosis
leukopenia
polycythemia
thrombocytopenia

_____ **1.** A deficiency in number of RBCs or decreased hemoglobin content of blood.

_____ **2.** An abnormal increase in RBCs.

_____ **3.** An abnormal increase in WBCs.

_____ **4.** A deficiency in platelets.

_____ **5.** A deficiency in WBC's.

C. White Blood Cell Structure and Characteristics

Fill in the blank with the correct answer. Terms may be used more than once.

basophils
eosinophils
leukocytes
lymphocytes
monocytes
neutrophils
PMNs
segs

_____ **1.** 60% to 70% of all WBCs.

_____ **2.** 2% to 4% of all WBCs.

_____ **3.** 0.5% to 1% of all WBCs.

_____ **4.** 20% to 25% of all WBCs.

_____ **5.** 3% to 8% of all WBCs.

_____ **6.** Nucleus with 2 to 5 connected lobes; pale lilac granules; larger than RBCs.

_____ **7.** Nucleus with 2 or 3 lobes; red-orange granules; larger than RBCs.

_____ **8.** Nucleus difficult to see; large deep blue-purple granules; larger than RBCs.

_____ **9.** Round nucleus that is dark purple; sky blue cytoplasm, no visible granules; about the same size as RBCs.

_____ **10.** Kidney-shaped nucleus; blue-gray cytoplasm, no visible granules; significantly larger than RBCs.

_____ **11.** Abbreviation for polymorphonuclear leukocytes.

_____ **12.** Nickname for neutrophils.

_____ **13.** General name for all of the WBCs.

D. ABO and Rh Blood Typing

Name the antigens present on the RBCs and the antibodies present in plasma of each blood type.

Antigens   **Antibodies**

1. O+   _____   _____

2. A−   _____   _____

3. B−   _____   _____

4. AB+   _____   _____

5. Based on what you know about antigens and antibodies, what blood type is the universal donor? _____ Explain.

6. What blood type is the universal recipient? _____ Explain.

E. Hematocrit

1. Define hematocrit.

2. Is a hematocrit of 47% within the normal range for a male?

3. Michaela is an athlete who has been training for several years only in Florida. She had an annual physical, and the doctor read her hematocrit as 59%. What could this reading potentially tell the doctor?

F. Application

1. Burke has a bacterial infection with inflammation. What type of WBC increases in number to help him fight the infection?

2. Staci had a differential WBC count that showed 60% neutrophils, 10% eosinophils, 1% basophils, 22% lymphocytes, and 7% monocytes. What could have caused this?

3. Jim's doctor found out that although Jim is Rh−, he received Rh+ blood for the first time. Will he have a transfusion reaction?

4. If Brittany is blood type O− and her husband is blood type AB−, what effect can the Rh factor have on their second child?

5. Valerie has type AB+ blood. What type(s) of blood can she receive in a transfusion?

21

STRUCTURE OF THE HEART

After completing this exercise, you should be able to:

- Describe the location and coverings of the heart and the three layers of the heart wall
- Identify major features of cardiac muscle tissue on a prepared microscope slide
- Identify the major heart structures on models or charts
- Describe the flow of a drop of blood through the pulmonary and systemic circulations, listing the vessels, chambers, and valves
- Describe the changes that take place in the heart after birth
- Identify the selected heart structures on a dissected sheep heart

MATERIALS

- human heart model
- human torso or chart showing the pulmonary and systemic circulations
- colored pencils
- preserved sheep heart (or other mammal heart)
- dissecting instruments, trays, disposable gloves, 5-inch blade knife
- compound microscope with cardiac muscle slide (for demonstration)

Your heart beats without external stimulation and rests only between heart beats. The heart is a small double pump that simultaneously pumps blood to and from body cells through the systemic circulation and to and from the lungs through the pulmonary circulation.

A. COVERINGS AND LAYERS OF THE HEART

The **heart** is about the size of a fist and lies in the thoracic cavity. The **base** of the heart is the wide superior portion of the heart from which the great vessels emerge, and the **apex** of the heart is the inferior end pointing to the left. The heart is tilted at an angle so that its inferior surface lies against the diaphragm with two-thirds of the heart to the left side of the sternum.

The heart is located between the lungs in the thoracic cavity and is surrounded and protected by the pericardium (*peri-* = around). The pericardium consists of an outer, tough fibrous pericardium and an inner, delicate serous pericardium. The **fibrous pericardium** attaches to the diaphragm and also to the great vessels of the heart. Like all serous membranes, the **serous pericardium** is a double membrane composed of an outer parietal layer and an inner visceral layer. Between these two layers is the **pericardial cavity** filled with **serous fluid.**

The wall of the heart has three layers: the outer epicardium (*epi-* = on, upon; *cardia* = heart), the middle myocardium (*myo-* = muscle), and the inner endocardium (*endo-* = within, inward). The **epicardium** is the *visceral layer of the pericardium*. The majority of the heart is **myocardium** or cardiac muscle tissue. The **endocardium** is a thin layer of endothelium deep to the myocardium that lines the chambers of the heart and the valves.

ACTIVITY 1 COVERINGS AND LAYERS OF THE HEART

1 Label the structures on Figure 21.1(a) and (b).

2 Pronounce each term as you write in the answer.

3 Using contrasting colors, color the layers of the heart and the pericardium.

4 Examine a slide of cardiac muscle tissue. Observe the striations, branching cardiac cells, nuclei, and intercalated discs. Refer to the photomicrograph of cardiac muscle tissue in Exercise 6 Tissues.

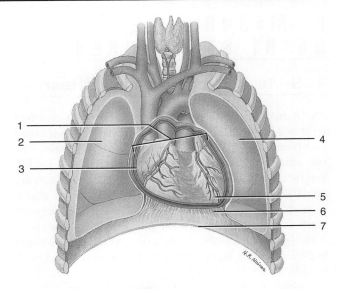

(a) Heart, anterior view

- apex of heart
- base of heart
- diaphragm
- fibrous pericardium (peri-CAR-dee-um)
- left lung
- right lung
- serous pericardium

1 _____

2 _____

3 _____

4 _____

5 _____

6 _____

7 _____

- endocardium
- fibrous pericardium
- myocardium
- parietal layer of serous pericardium
- pericardial cavity
- visceral layer of serous pericardium (epicardium)

8 _____

9 _____

10 _____

11 _____

12 _____

13 _____

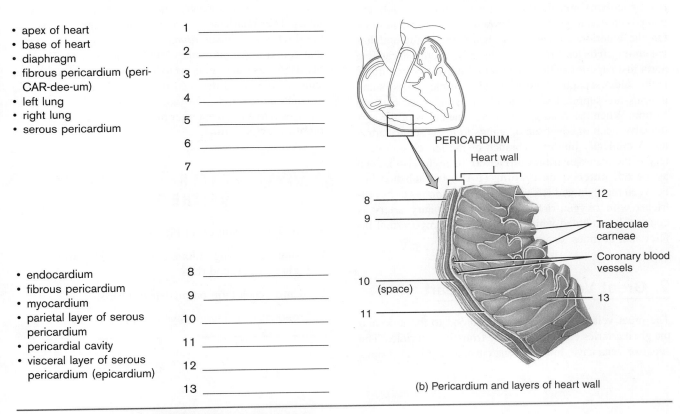

(b) Pericardium and layers of heart wall

FIGURE 21.1   Layers of the heart and pericardium.

B. MAJOR HEART STRUCTURES

1. Surface Structures of the Heart

The human heart has four chambers and is divided into right and left sides. Each side has an upper chamber called an **atrium** and a lower chamber called a **ventricle.** The two atria form the base of the heart and the tip of the left ventricle forms the apex. **Auricles** (*auricle* = little ear) are pouch-like extensions of the atria with wrinkled edges. Shallow grooves called **sulci** (sulcus, singular) externally mark the boundaries between the four heart chambers. Although a considerable amount of external adipose tissue is present on the heart surface for cushioning, most heart models do not show this.

Cardiac muscle tissue that composes the heart walls has its own blood supply and circulation, the **coronary** (*corona* = crown) **circulation.** Coronary blood vessels encompass the heart similar to a crown and are found in sulci. On the anterior surface of the heart, the **right** and **left coronary arteries** branch off the base of the ascending aorta just superior to the aortic semilunar valve, and travel in the sulcus separating the atria and ventricles. These small arteries are supplied with blood when the ventricles are resting. When the ventricles contract, the cusps of the aortic valve open to cover the openings to the coronary arteries. A clinically important branch of the left coronary artery is the **anterior interventricular branch,** also known as the **left anterior descending (LAD) branch** that lies between the right and left ventricles and supplies both ventricles with oxygen-rich blood. This coronary artery is commonly occluded which can result in a myocardial infarct and, at times, death.

2. Great Vessels of the Heart

The great veins of the heart return blood to the atria and the great arteries carry blood away from the ventricles. The superior vena cava, inferior vena cava, and coronary sinus return oxygen-poor blood to the right atrium. The **superior vena cava** returns blood from the head, neck, and arms; the **inferior vena cava** returns blood from the body inferior to the heart. The **coronary sinus** is a smaller vein that returns blood from the coronary circulation. Blood leaves the right atrium to enter the right ventricle. From here, oxygen-poor blood passes out the **pulmonary trunk,** the only vessel that removes blood from the right ventricle. This large artery divides into the **right** and **left pulmonary arteries** that carry blood to the lungs where it is oxygenated. Oxygen-rich blood returns to the left atrium through two **right** and two **left pulmonary veins.** The blood then passes into the left ventricle that pumps blood into the large **aorta.** The aorta distributes blood to the systemic circulation. The aorta begins as a short **ascending aorta,** curves to the left to form the **aortic arch,** descends posteriorly and continues as the **descending aorta.**

In fetal life, oxygen is obtained through the placenta from the mother and not from the lungs. Therefore, it is not detrimental to the baby's health for blood to bypass the lungs. The fetal heart contains a short, temporary vessel, the **ductus arteriosus** (*ductus* = duct; *arteria* = artery), that connects the pulmonary trunk and the aorta. The ductus arteriosus is a right heart to left heart shunt that reroutes some of the blood destined for the lungs to the systemic circulation through the aorta. The ductus arteriosus changes into a ligament after birth and remains as the **ligamentum arteriosum.**

ACTIVITY 2 EXTERNAL STRUCTURES OF THE HEART

1 Label the structures on Figure 21.2(a) and (b).

2 Using contrasting colors, color the heart structures in Figure 21.2(a) and (b).

3 Identify each term on models or charts.

4 Pronounce each term as you point to it.

- aortic arch
- ascending aorta
- auricle of left atrium
- auricle of right atrium
- branch of left coronary artery or LAD
- descending aorta
- inferior vena cava
- left pulmonary artery
- left pulmonary veins

- left ventricle
- ligamentum arteriosum
- pulmonary trunk
- right coronary artery
- right pulmonary artery
- right pulmonary veins
- right ventricle
- superior vena cava

1 _____
2 _____
3 _____
4 _____
5 _____
6 _____
7 _____
8 _____
9 _____
10 _____
11 _____
12 _____
13 _____
14 _____
15 _____
16 _____
17 _____

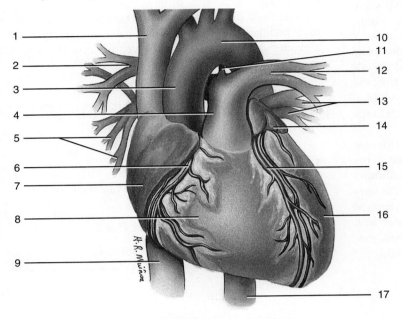

(a) Anterior view showing surface features

- aortic arch
- ascending aorta
- coronary sinus
- inferior vena cava
- left pulmonary artery

- left pulmonary veins
- ligamentum arteriosum
- right pulmonary artery
- right pulmonary veins
- superior vena cava

18 _____
19 _____
20 _____
21 _____
22 _____
23 _____
24 _____
25 _____
26 _____
27 _____

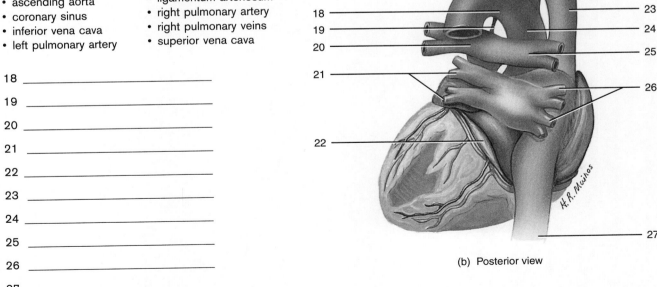

(b) Posterior view

FIGURE 21.2   **External structure of the heart.**

3. Internal Structures of the Heart

The heart has four valves that control the one-way flow of blood: two **atrioventricular (AV) valves** and two **semilunar valves** (*semi-* = half; *lunar* = moon). Blood passing between the right atrium and the right ventricle goes through the right AV valve, the **tricuspid valve** (*tri* = three; *cusp* = flap). The left AV valve, the **bicuspid valve,** is between the left atrium and the left ventricle. This valve clinically is called the **mitral valve** (*miter* = tall, liturgical headdress) because the open valve resembles a bishop's headdress. String-like cords called **chordae tendineae** (tendinous strands) attach and secure the cusps of the AV valves to enlarged **papillary muscles** that project from the ventricular walls. Chordae tendinae allow the AV valves to close during ventricular contraction, but prevent their cusps from getting pushed up into the atria.

The two semilunar valves allow blood to flow from the ventricles to great arteries and exit the heart. Blood in the right ventricle goes through the **pulmonary (semilunar) valve** to enter the pulmonary trunk, a large artery. The **aortic (semilunar) valve** is located between the left ventricle and the aorta. These two semilunar valves are identical, with each having three pockets that fill with blood, preventing blood from flowing back into the ventricles.

The two ventricles have a thick wall between them called the **interventricular septum.** Between the two atria is a thinner **interatrial septum.** In the fetus, there is a hole in the interatrial septum called the foramen ovale. The **foramen ovale** allows blood to bypass the lungs and go from the right atrium to the left atrium, forming another right heart to left heart shunt. The **fossa ovalis,** a connective tissue membrane remnant, forms over and closes the fetal foramen ovale after birth.

Note the difference in thickness between the right ventricle and the left ventricle. Which one has to pump blood a greater distance and therefore has to pump blood with more force?

ACTIVITY 3 INTERNAL FEATURES OF THE HEART

1 Label the structures on Figure 21.3.

2 Using contrasting colors, color the internal structures of the heart in Figure 21.3.

3 Identify each term on models or charts.

4 Pronounce each term as you point to it.

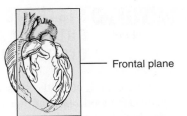

Frontal plane

- aortic (semilunar) valve
- bicuspid valve (mitral)
- chordae tendineae (CHOR-dee ten-DIN-ee)
- coronary sinus opening
- interventricular septum (inter-ven-TRIC-u-lar)
- left atrium
- left ventricle
- papillary muscle (PAP-ih-lary)
- pulmonary (semilunar) valve
- right atrium
- right ventricle
- tricuspid valve

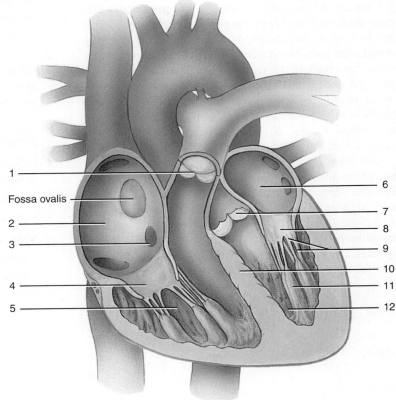

Fossa ovalis

Heart, frontal section

1 _____

2 _____

3 _____

4 _____

5 _____   8 _____   11 _____

6 _____   9 _____   12 _____

7 _____   10 _____

FIGURE 21.3   Internal structures of the heart.

C. SYSTEMIC AND PULMONARY CIRCULATIONS

As you trace a drop of blood through the heart to the lungs and then to the rest of the body, you will be examining the pulmonary and systemic circulations. The **pulmonary circulation** takes blood from the right ventricle to the lungs and back to the left atrium. Simultaneously, the **systemic circulation** takes blood from the left ventricle to the body tissues and back to the right atrium. Each circulation begins and ends at the heart, and each circulation is composed of arteries, capillaries, and veins. Arteries carry blood from the heart to the capillaries, microscopic vessels within tissues. Blood travels from capillaries into veins, which carry blood back to the heart.

The thinner walled atria receive blood returning to the heart from the great veins. Both atria contract simultaneously to pump the blood into the ventricles. The larger, thick ventricular walls are double pumps that contract simultaneously to send the blood from the right ventricle to the pulmonary circulation and from the left ventricle to the systemic circulation. The wall of the left ventricle is thicker than the right because the left side requires more force to pump blood through the systemic circulation.

ACTIVITY 4 SYSTEMIC AND PULMONARY CIRCULATIONS

1 In Figure 21.4, color the vessels that are carrying oxygen-poor blood *blue* and the vessels carrying oxygen-rich blood *red*, being careful to note the color switch in the pulmonary vessels. Color the four capillary beds *purple*.

2 Trace the pathway of blood on Figure 21.4 through the pulmonary circulation with one color of arrows, and the systemic circulation with different colored arrows, starting and ending with the right atrium.

3 Using a heart model, trace the pathway of blood through the great vessels and heart structures.

4 Indicate whether the following blood vessels contain oxygen-poor or oxygen-rich blood.
- aorta
- pulmonary arteries
- pulmonary trunk
- pulmonary veins
- venae cavae

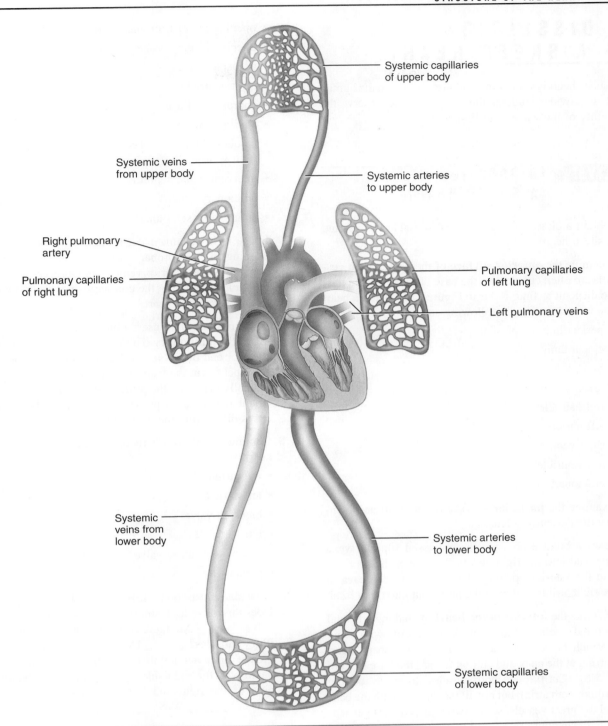

Systemic capillaries
of upper body

Systemic veins
from upper body

Systemic arteries
to upper body

Right pulmonary
artery

Pulmonary capillaries
of left lung

Pulmonary capillaries
of right lung

Left pulmonary veins

Systemic
veins from
lower body

Systemic arteries
to lower body

Systemic capillaries
of lower body

FIGURE 21.4   **Systemic and pulmonary circulations.**

E. DISSECTION
OF A SHEEP HEART

The sheep heart is similar in structure and size to the human heart. It provides students the opportunity to observe the flexibility of the valves and tissues.

**ACTIVITY 5 DISSECTION OF
A SHEEP HEART**

1 Obtain a dissecting tray, tools, disposable gloves, and a sheep heart.

2 Examine the **anterior surface** of the heart. Great vessels are often cut close to the base of the heart and may be difficult to find. Refer to Figure 21.5(a) and a heart model to identify the following structures:
- pericardium (if present)
- epicardium
- base
- apex
- right auricle
- left auricle
- right ventricle
- left ventricle
- pulmonary trunk

3 Examine the **posterior surface** of the heart and identify the coronary sulcus.

4 Insert a blunt probe into the collapsed superior vena cava and into the right atrium. Maneuver the probe to find the interior opening of the inferior vena cava in the right atrium and push the probe out into this vessel.

5 Examine the **interior of the heart** by making a frontal (coronal) section of the heart. Using a knife with about a 5-inch blade, make a coronal cut of a sheep heart starting at the apex and cutting toward the base (Figure 21.5(b)). Cut through both auricles (to ensure cutting through both atria), but not all the way through the base and the great vessels, so the two halves do not get separated. This cut allows you to observe both atria and both ventricles simultaneously (similar to a heart model), and easily compare the size of the walls of the right and left ventricles.

6 Using Figure 21.5(b), identify the following interior structures on the right side of the heart:
- myocardium
- endocardium
- right atrium
- right auricle

- opening of superior and inferior vena cava
- opening of the coronary sinus
- tricuspid valve
- right ventricle
- chordae tendinae
- papillary muscles
- moderator band (cord between the two walls of the right ventricle)
- interventricular septum
- pulmonary trunk
- pulmonary (semilunar) valve

7 In the right atrium, insert a blunt probe in the small opening of the coronary sinus that is medial to the opening of the inferior vena cava. Observe the movement of the probe in the coronary sinus from the posterior view of the heart.

8 In the right atrium, insert a blunt probe into the opening of the **pulmonary trunk** and push it through to the superior end of the vessel. Remove the probe and use a scalpel to cut the wall of the pulmonary trunk longitudinally to expose the **pulmonary (semilunar) valve.** Count the three cusps. How does this valve and its cusps differ from the tricuspid valve?

9 Continue identifying structures on the left side of the heart.
- left atrium
- left auricle
- bicuspid valve
- left ventricle
- aortic (semilunar) valve
- aorta

10 How many cusps does the bicuspid valve have? _____ Does this valve look similar otherwise to the other AV valve? _____ Are there **chordae tendineae** and **papillary muscles?** _____ Does the left ventricle have a greater or lesser number of papillary muscles compared with the right side? _____ Compare the thickness of the right and left ventricles. Which one is thicker? _____ Why?_____

11 Look just above the cusps of the aortic valve for the **openings to the right** and **left coronary arteries.** Use the blunt probe to push into these small vessels.

12 Dispose of any removed dissection material in the proper container (*not* the sink!).

13 Wash your dissection pan, instruments, and hands with soap and water when finished.

14 Clean up your lab space and wash the countertops with disinfectant.

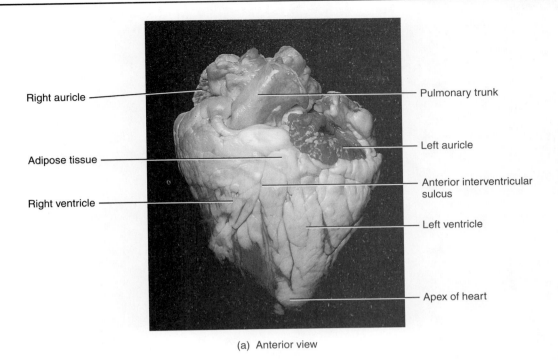

Right auricle

Adipose tissue

Right ventricle

Pulmonary trunk

Left auricle

Anterior interventricular sulcus

Left ventricle

Apex of heart

(a) Anterior view

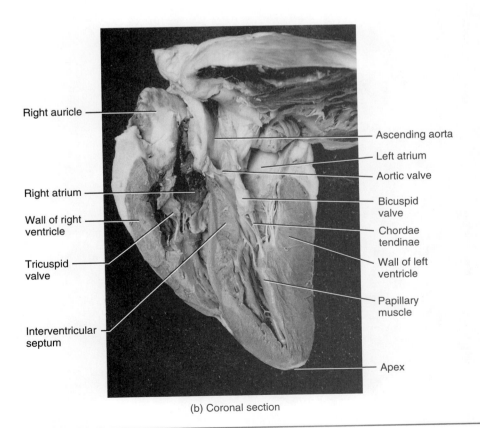

Right auricle

Right atrium

Wall of right ventricle

Tricuspid valve

Interventricular septum

Ascending aorta

Left atrium

Aortic valve

Bicuspid valve

Chordae tendinae

Wall of left ventricle

Papillary muscle

Apex

(b) Coronal section

FIGURE 21.5   Sheep heart.

REVIEWING YOUR KNOWLEDGE

A. Major Heart Structures

Write the name of the structure that each phrase describes.

apex
atria
auricles
base
chordae tendinae
coronary arteries
endocardium
epicardium
ligamentum arteriosum
myocardium
papillary muscles
ventricles

_____ **1.** Arteries that supply blood to cardiac muscle.

_____ **2.** Layer of heart wall containing cardiac muscle.

_____ **3.** Wrinkled extensions of the atria.

_____ **4.** Lines the heart chambers.

_____ **5.** Pointed inferior part of the heart.

_____ **6.** Two heart pumps; lower heart chambers.

_____ **7.** Superior heart chambers.

_____ **8.** Another name for visceral pericardium.

_____ **9.** Wide, superior part of the heart.

_____ **10.** Enlarged muscles in ventricles attached to chordae tendinae.

_____ **11.** Strings attached to AV cusps.

_____ **12.** Remnant of a fetal vessel that closed at birth.

B. The Heart and Pulmonary Circulation

Trace a drop of blood from the *neck area* to the heart, to the lungs, and out of the heart to the systemic circulation by placing the following structures in order, numbering them 1–14.

_____ aorta

_____ aortic (semilunar) valve

_____ bicuspid valve (mitral)

_____ left atrium

_____ left ventricle

_____ pulmonary arteries

_____ pulmonary capillaries

_____ pulmonary (semilunar) valve

_____ pulmonary trunk

_____ pulmonary veins

_____ right atrium

_____ right ventricle

_____ superior vena cava

_____ tricuspid valve

HEART SOUNDS, PULSE RATE, AND BLOOD PRESSURE

OBJECTIVES

After completing this exercise, you should be able to:

- Describe each heart sound and state when each one occurs
- Describe the relationship of auscultated heart sounds, pulse rate, and heart rate
- Identify the events responsible for systolic and diastolic pressure
- Measure systolic and diastolic blood pressure at rest and after exercise
- Discuss how exercise and body position affect blood pressure

MATERIALS

- stethoscope and alcohol swabs
- sphygmomanometer
- stop watch or watch with second hand
- meter stick

A. HEART SOUNDS, HEART RATE, AND PULSE RATE

Auscultation (*auscult-* = listening) means listening to body sounds, typically using a stethoscope. Two particular **heart sounds** can be detected during one heartbeat. These sounds occur after the heart valves quietly close and blood strikes against the closed valve, causing turbulence that we can hear with a stethoscope. Although there are four sounds generated during one heartbeat, only the first and second sound can easily be heard without additional amplification. The first sound, **lubb,** is a little longer and louder than **dupp** (or **dubb**), the second

sound that occurs shortly after the first. The first sound (S1) is due to blood turbulence that occurs when the two atrioventricular (AV) valves close during ventricular **systole,** when the ventricles are contracting. The second sound (S2) is at ventricular **diastole,** when the ventricles relax and the two semilunar valves close. The sounds that you will hear are lubb-dupp, pause. . .lubb-dupp, pause. Remember that these two heart sounds, lubb-dupp, equal one heartbeat. The number of heartbeats/minute is the **heart rate.**

When the ventricles contract, a blood pressure wave is produced that travels in the arteries and can be felt as your **pulse.** The number of heartbeats per minute (heart rate) will be very close to, but not necessarily equal to, the number of pulses per minute.

CLINICAL NOTE: *Heart murmurs are abnormal heart sounds caused by the restriction of blood flow or backflow of blood that typically indicate an abnormality in valve structure or function. It usually takes special training and a very quiet room to be able to detect heart murmurs.*

ACTIVITY 1 HEART SOUNDS, HEART RATE, AND PULSE RATE

1 Auscultate the heart sounds.

- Obtain a stethoscope and alcohol swabs.
- Clean the earpieces with the alcohol swab and let air dry.

- Earpieces should be pointed forward when placed in the external ear canal. This coincides with the ear contour for comfort and also makes listening easier.
- Gently tap the bell with the earpieces in place to check that you can hear sounds before auscultation.
- Auscultation of the aortic valve may be performed on your own chest or that of your lab partner. Palpate the suprasternal (jugular) notch at the superior end of the sternum. Then palpate the sternal angle, a raised area where the manubrium and sternal body articulate. Now drop down another 1 inch on the sternum from the sternal angle and then move right 1 inch. You should be in the 2nd intercostal space just to the right of the sternum (Figure 22.1). Use the large bell end of the stethoscope to hear the heart sounds.

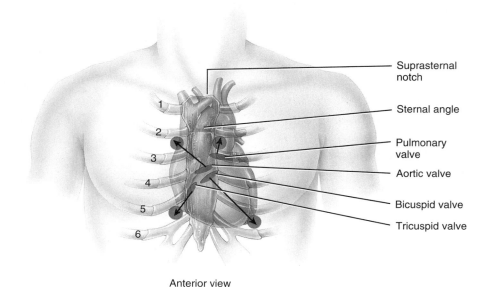

Anterior view

FIGURE 22.1 **Locations for auscultation of the heart.**

2 Calculate your heart rate at rest.

- Count the number of heartbeats in 15 seconds and record your answer in Table 22.1.

- Calculate beats/minute by multiplying this number by four and record. The number of heartbeats/minute is called your **heart rate.** Estimate the time between heartbeats and record.

3 Calculate your pulse rate at rest using the radial or carotid artery.

Radial Pulse

- Locate the groove in your partner's wrist between the radius and the ligament just medial to the radius. Palpate the radial artery by pressing down with your index and middle fingers. Because the thumb has a noticeable pulse of its own, it is not used for this activity.

- Count the number of pulses per 15 seconds and record in Table 22.1. Multiply this number by 4 to calculate the **pulse rate** and record.

Carotid Pulse

- Locate your lab partner's Adam's apple (thyroid cartilage of larynx). Using your index and middle fingers again, palpate the carotid artery on either side of the larynx.

- Count the number of pulses per 15 seconds and record in Table 22.1. Multiply this number by 4 to calculate the pulse rate and record.

4 Calculate your heart rate and pulse rate after exercise.

- If your lab partner does not have heart problems, have him or her run in place or briskly walk up and down stairs for 1 minute.

- Take your lab partner's carotid pulse immediately. Count the number of pulses in 15 seconds, and record in Table 22.1. Calculate the pulse rate/minute.

- Listen to your lab partner's heartbeat immediately and count the number of beats/15 seconds and record in Table 22.1. Calculate the heart rate/minute and record. Estimate the time between heartbeats and record.

- Repeat taking the pulse every 30 seconds until the pulse rate returns to the initial resting rate.

- Record the time for recovery: _____ minutes.

5 Answer the discussion questions with your lab group.

❓ DISCUSSION QUESTIONS: HEART SOUNDS, HEART RATE, AND PULSE RATE

1 At rest, which heart sound is louder, lubb or dupp? Explain.

2 How did your pulse rate/minute compare with your heart rate/minute?

3 Compare the length of the pause between heartbeats after exercise with the pre-exercise length.

4 Compare pulse rate recovery time with other lab groups. Hypothesize with your group what could cause the different recovery times.

TABLE 22.1 Heart Rate and Pulse Rate Before and After Exercise

ACTIVITY	BEATS IN 15 SEC	BEATS/MIN	SEC BETWEEN BEATS
Heart rate at rest			
Heart rate after exercise			
Radial pulse rate at rest			
Radial pulse rate after exercise			
Carotid pulse rate at rest			
Carotid pulse rate after exercise			

B. BLOOD PRESSURE

Blood pressure, the pressure exerted by blood against blood vessel walls, is highest in the aorta and large elastic arteries, and decreases as the arteries branch and blood travels further from the heart. Blood pressure drops significantly in the arterioles (small arteries) and steadily decreases through capillaries, venules (small veins), and veins, and drops to zero in the right atrium. The difference in blood pressure between two areas of the circulatory system is the **blood pressure gradient.** The blood pressure gradient between the aorta and right atrium causes blood to flow through the systemic circulation, and the blood pressure gradient between the pulmonary trunk and the left atrium causes blood to flow through the pulmonary circulation.

With each ventricular contraction, blood pressure fluctuates in the large arteries (i.e., the aorta, pulmonary trunk, and other elastic and large muscular arteries). Blood pressure during ventricular systole, **systolic blood pressure,** is higher than blood pressure during ventricular diastole, **diastolic blood pressure.** These fluctuations diminish within the arterioles, and little or no fluctuations in blood pressure between systole and diastole are observed in the capillaries and venous vessels.

The blood pressure gradient between veins and the right atrium is low, making it difficult for venous blood flow to overcome the force of gravity. Therefore, when standing with arms hanging down, blood may pool in the large veins of the limbs.

1. Blood Pressure Measurements

Arterial blood pressure is reported in millimeters of mercury (mm Hg) and can be measured directly by inserting a pressure transducer into a large artery or indirectly with a sphygmomanometer. A **sphygmomanometer** (SFIG-moe-mah-NOH-meh-ter), or blood pressure cuff, can be used to measure systolic and diastolic blood pressure in any large artery. Clinically, the brachial artery is used most often. The sphygmomanometer contains a pressure gauge attached to an inflatable rubber cuff that is connected by a rubber tube to a hand pump (rubber bulb) or automatic pump. Clinically, it is important to use the right size cuff—pediatric, standard, or large arm—to obtain an accurate reading. The pump is used to inflate the rubber cuff to a pressure greater than the systolic pressure. This puts pressure on the artery, flattens it, and stops blood flow in the artery. A stethoscope is placed over the brachial artery in the antecubital area, and the pressure is slowly released by opening a valve. When blood pressure is greater than the pressure in the cuff, the artery opens and blood flow returns. The examiner listens for the sound caused by the turbulent flow of blood that can be heard until blood flow returns to normal. These sounds are called the **Korotkoff**

(koh-ROT-koff) **sounds.** Although there is a range of values considered normal, the average normal systolic pressure (the first sound heard) is 120 mm Hg, and normal diastolic pressure (the last, faint sounds heard) is 80 mm Hg, usually written as 120/80.

Venous blood pressure can be measured directly with a pressure transducer inserted into a venous vessel or indirectly. The indirect measurement of venous pressure described in the following activity is an estimate at best. Blood pressure in veins is very low. The average venous blood pressure is 16 mm Hg.

ACTIVITY 2 BLOOD PRESSURE MEASUREMENTS

1 Measure resting systemic arterial blood pressure.

- Work in teams of two, alternating who will be the subject and who will measure blood pressure.

- The subject should sit and rest for 5 minutes before measuring blood pressure.

- Wipe the earpieces of the stethoscope with alcohol and completely deflate the blood pressure cuff.

- Place the cuff of the sphygmomanometer around the arm as shown in Figure 22.2. If the cuff does not fit the arm, the blood pressure measurements will probably be inaccurate. The inflatable portion of the cuff should be over the anterior surface of the arm. If there is an arrow on the cuff, place it over the brachial artery. The bottom of the cuff should be approximately one inch above the elbow.

- Close the valve on the rubber bulb.

- Place the large bell of the stethoscope over the brachial artery and insert the earpieces. The brachial artery is located in the anterior arm, immediately superior to the antecubital fossa and lateral to the biceps brachii.

- Inflate the cuff to 160–180 mm Hg.

- Immediately use the valve on the hand pump to slowly release air to deflate the cuff, and listen for the first sound.

- The first sound, the systolic pressure, is the return of blood flow through the partially occluded brachial artery. Watch the pressure gauge and continue to listen; the sound will increase, then muffle, and stop. The diastolic pressure is the pressure when the last, faint sound is heard.

- Record blood pressure measurements in Table 22.2.

- Deflate the cuff and have subject rest for 2 minutes before repeating the blood pressure measurement.

2 Record second value in Table 22.2.

3 As a class, discuss the results.

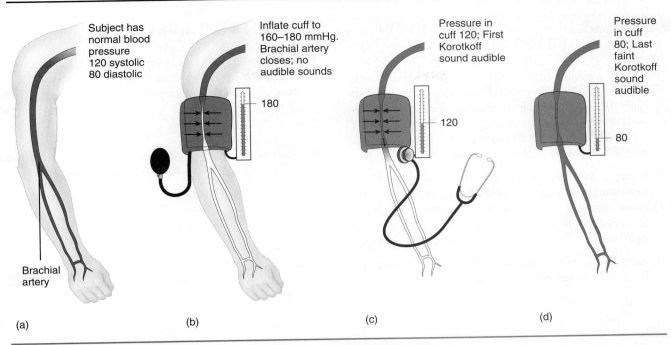

FIGURE 22.2 **Measuring arterial blood pressure with a sphygmomanometer.**

TABLE 22.2 Resting Blood Pressures		
SUBJECT	**SYSTOLIC PRESSURE**	**DIASTOLIC PRESSURE**
Subject 1	1.	1.
	2.	2.
	Avg.	Avg.
Subject 2	1.	1.
	2.	2.
	Avg.	Avg.

2. Regulation of Blood Pressure

The body maintains blood pressure to ensure adequate blood flow to body tissues (tissue perfusion). If blood pressure is too low, tissue perfusion may not be sufficient to provide oxygen and nutrients to cells, especially brain neurons, or to remove metabolic wastes. If blood pressure is too high, it may cause damage to capillaries and the endothelial lining of blood vessels. Eye exams can often detect early hypertension. Damaged retinal vessels can be observed with an ophthalmoscope.

Systemic blood pressure (BP) increases when cardiac output increases and when resistance to blood flow increases. Blood viscosity, total blood vessel length, and blood vessel diameter affect resistance to blood flow. However, blood vessel diameter is the only variable that can be changed to make immediate changes in blood pressure. Increasing blood vessel diameter (vasodilation) decreases resistance and lowers BP, while decreasing blood vessel diameter (vasoconstriction) increases resistance and elevates BP.

ACTIVITY 3 REGULATION OF BLOOD PRESSURE

1 Choose a subject from your lab group. Observe the effect of *body position* on blood pressure (BP) and heart rate (HR) by measuring BP and HR in the supine position, immediately after standing, and after standing for 2 minutes. Record values in Table 22.3. Pool class data, calculate averages, and graph the data.

2 Observe the effect of *exercise* on blood pressure by measuring systolic and diastolic BP and HR before exercise (standing, at rest) and after 5 minutes of exercise (e.g., running in place, climbing stairs). Record values in Table 22.4. Pool class data and calculate averages.

3 Answer discussion questions with your lab partners. Refer to your textbook when necessary.

? DISCUSSION QUESTIONS

1 When was BP lowest, in the supine or standing position? Why was BP lower in this position? Why was BP changed when the position was changed?

2 Describe the change in BP and heart rate that occurred with exercise. Explain how the body changed BP.

TABLE 22.3 Effect of Body Position on Blood Pressure

SUBJECT	SUPINE BP AND HR	BP AND HR IMMEDIATELY AFTER STANDING	BP AND HR AFTER STANDING FOR 2 MINUTES
Subject 1	Systolic BP = Diastolic BP = HR =	Systolic BP = Diastolic BP = HR =	Systolic BP = Diastolic BP = HR =
Subject 2	Systolic BP = Diastolic BP = HR =	Systolic BP = Diastolic BP = HR =	Systolic BP = Diastolic BP = HR =
Class Average	Systolic BP = Diastolic BP = HR =	Systolic BP = Diastolic BP = HR =	Systolic BP = Diastolic BP = HR =

TABLE 22.4 Effect of Exercise on Blood Pressure

SUBJECT	RESTING BP	POSTEXERCISE BP
Subject 1	Systolic BP = Diastolic BP = HR =	Systolic BP = Diastolic BP = HR =
Subject 2	Systolic BP = Diastolic BP = HR =	Systolic BP = Diastolic BP = HR =
Class Average	Systolic BP = Diastolic BP = HR =	Systolic BP = Diastolic BP = HR =

REVIEWING YOUR KNOWLEDGE

A. Heart Sounds

Fill in the blank with the correct term.

1. The first heart sound heard is _____ that is due to blood hitting against the _____ valves.

2. The second heart sound heard is _____ that is due to blood hitting against the _____ valves.

3. What is a heart murmur?

4. Which heart sound is the loudest sound when auscultated, the lubb or dupp?

B. Blood Pressure

Write the correct term for the following description.

_____ **1.** Term given for arterial pressure during ventricular systole.

_____ **2.** Term given for arterial pressure during ventricular diastole.

_____ **3.** Device used to measure arterial blood pressure in the brachial artery.

_____ **4.** Average normal adult arterial blood pressure.

_____ **5.** Sounds of turbulent blood flow that occur when blood flow resumes in an artery that has been occluded.

C. Regulation of Blood Pressure and Blood Flow

Mark the following True or False. If False, underline and change the word(s) to make the statement True.

_____ **1.** Increasing heart rate increases blood pressure.

_____ **2.** Systemic vasoconstriction decreases blood pressure.

_____ **3.** Increasing arterial blood pressure increases blood flow.

_____ **4.** Vasoconstriction of the renal arteries (arteries supplying blood to kidneys) would decrease blood flow to the kidneys.

_____ **5.** Blood pressure is higher in the supine position than in the standing position.

_____ **6.** Blood pressure decreases during exercise.

BLOOD VESSEL STRUCTURE AND IDENTIFICATION

23

After completing this exercise, you should be able to:

- Compare and contrast the structure of arteries, capillaries, and veins

- Identify the major arteries and veins of the systemic circulation on models or charts

- Identify the cerebral arterial circle (circle of Willis) on models or charts and describe its function

- Identify the major vessels of the pulmonary circulation on models or charts

- Identify the major vessels of fetal circulation and describe changes in fetal circulation after birth

- Trace blood flow through the cardiovascular system

- models or charts of human arteries and veins

Arteries carry blood away from the heart and divide into smaller vessels called **arterioles** that branch into tiny **capillaries.** An exchange of nutrients and gases occurs between blood and interstitial fluid (fluid surrounding tissue cells) at the capillary level. Capillaries join to form small **venules** that will merge to form larger **veins** that carry blood back to the heart. The structure of these vessels reflects their functions.

There are two circulatory routes within the body: the pulmonary circulation and the systemic circulation. In the **systemic circulation,** arteries carry oxygen-rich blood to body tissues, and veins return oxygen-poor blood to the heart where it will then enter the pulmonary circulation. In the **pulmonary circulation,** arteries carry oxygen-poor blood from the right ventricle to the lungs where gas exchange occurs, and veins return oxygen-rich blood back to the left atrium.

A. BLOOD VESSEL STRUCTURE

1. Arteries

Arterial walls have three layers that enclose the center space or **lumen** where the blood flows. The **outer layer** is composed mainly of elastic and collagen fibers (proteins) that provide support and protection. The **middle** and thickest **layer** contains elastic fibers and smooth muscle fibers (cells) encircling the diameter of the vessel. Contraction of smooth muscle fibers causes a decrease in lumen diameter or **vasoconstriction,** whereas relaxation causes **vasodilation,** an increase in lumen diameter. Elastic fibers allow the vessel to be stretched and return to its original shape. The **inner layer** of the arterial wall contains simple squamous epithelium (**endothelium**), a basement membrane, and elastic tissue.

Arteries branch into smaller and smaller arteries that eventually form **arterioles** (little arteries), small blood vessels from which capillaries branch. Arterioles play a major role in controlling blood pressure and control blood flow into capillaries.

2. Capillaries

Capillaries have the smallest diameter and thinnest walls of any blood vessels. Their lumen is so small that red blood cells can only pass through one at a time. Capillary walls are composed of a single layer of endothelial cells (simple squamous epithelium) supported by a basement membrane. The exchange of substances between the blood and the tissues occurs only at the capillary level due to their thin walls and special structural permeability.

3. Veins

Blood flows from capillaries into **venules** (little veins) that drain into veins. The walls of **veins** also contain three layers. Compared with arterial vessels, the walls of veins are thinner and contain fewer smooth muscle fibers and elastic fibers. The **lumen** of veins is larger than arteries and often appears collapsed in tissue sections. Veins contain approximately 60% of the total blood volume and are called blood reservoirs.

The blood pressure gradient in the veins is very small and often is not enough to overcome gravity. Muscular activity squeezes the veins and pushes blood toward the heart while **valves,** found in many veins, prevent the backflow of blood, especially in the limbs.

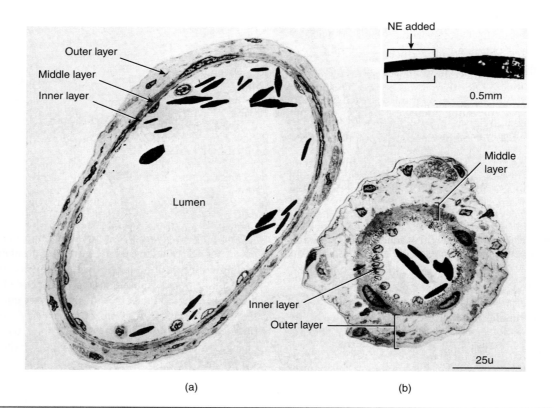

(a)   (b)

FIGURE 23.1   Transverse sections through a frog artery.

ACTIVITY 1 STRUCTURE OF BLOOD VESSELS

1 In Figure 23.1, compare lumen diameter and thickness of an arterial wall in two sections of a frog arteriole. Norepinephrine (NE) was applied to one area of the arteriole to induce vasoconstriction. Which cross-section is vasoconstricted, Figure 23.1(a) or (b) ___?

2 Examine the blood vessels in Figure 23.2. Identify the arteriole and a capillary. *Hint:* Look at the size of the vessel and the blood cells in the lumen.

3 Identify and label the layers of the blood vessels in Figure 23.3.

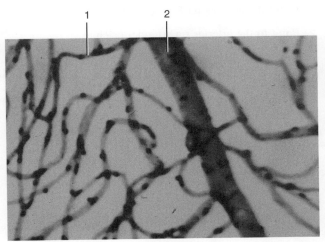

250×

- arteriole
- capillary

1 _____

2 _____

FIGURE 23.2   Photomicrograph of capillary network.

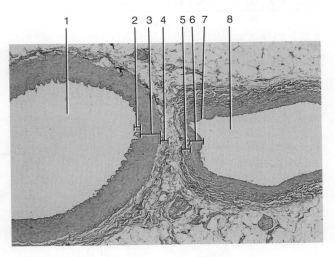

- external layer of artery
- external layer of vein
- internal layer of artery
- internal layer of vein
- lumen of artery
- lumen of vein
- middle layer of artery
- middle layer of vein

1 _____

2 _____

3 _____

4 _____

5 _____

6 _____

7 _____

8 _____

FIGURE 23.3   Transverse section through small artery and vein.

B. SYSTEMIC ARTERIES

All major arteries of the systemic circulation branch off the aorta. These arteries transport blood to different regions of the body where blood is distributed to smaller arteries and finally to arterioles that connect to capillaries.

1. Major Arteries of the Ascending Aorta and Aortic Arch

Blood is ejected from the left ventricle into the **ascending aorta.** The **right** and **left coronary arteries** branch off the ascending aorta near its origin and travel to the heart. The ascending aorta curves superiorly toward the left and becomes

the **aortic arch.** Three major arteries branch off the aortic arch in the following order from right to left: the **brachiocephalic trunk,** the **left common carotid artery,** and the **left subclavian artery.** The brachiocephalic trunk divides to form the **right common carotid** and **right subclavian arteries.**

ACTIVITY 2 MAJOR ARTERIES OF THE ASCENDING AORTA AND AORTIC ARCH

1 Label the arteries on Figure 23.4.

2 Point to the arteries on a model or chart.

3 Pronounce the names as you point to them.

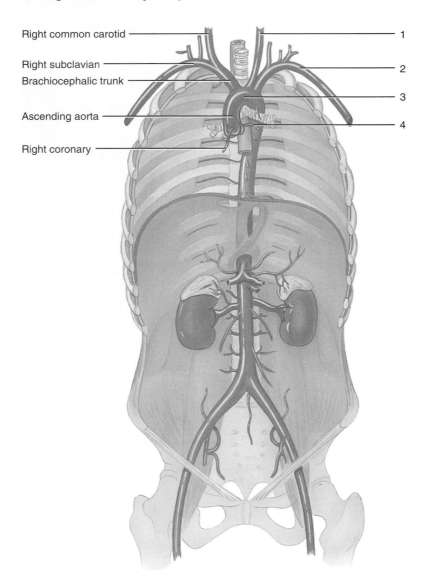

Right common carotid — 1
Right subclavian —
Brachiocephalic trunk — 2
— 3
Ascending aorta — 4
Right coronary —

- aortic arch
- left common carotid
- left coronary
- left subclavian

1 _____

2 _____

3 _____

4 _____

FIGURE 23.4   **Major arteries of the ascending aorta and aortic arch.**

2. Major Arteries Supplying the Head (Including the Circle of Willis)

The common carotid arteries branch to form the internal and external carotid arteries. The **internal carotid arteries** travel through the neck and enter the skull through the carotid canal and supply blood to the anterior and middle brain areas. The **external carotid arteries** ascend along the lateral surface of the neck and terminate as two arteries near the temporomandibular joint to supply structures external to the skull.

There are two **vertebral arteries,** one branching off each subclavian artery. The vertebral arteries travel superiorly through the transverse foramen of the cervical vertebrae and the foramen magnum to supply blood to the posterior brain areas. The paired internal carotid and paired vertebral arteries supply blood to the **cerebral arterial cir-**cle (circle of Willis), a ring of blood vessels formed by the convergence of several arteries. Blood supply to the brain is vital, and the configuration of the arteries supplying the brain ensures continuous blood flow even when blockage occurs in a blood vessel.

ACTIVITY 3 MAJOR ARTERIES SUPPLYING THE HEAD

1 Label the arteries supplying blood to the cerebral arterial circle (circle of Willis) in Figure 23.5(a) and (b).

2 Identify the arteries on models or charts.

3 Pronounce the names as you point to them.

4 Palpate your right common carotid artery pulse by placing your fingers just lateral to your larynx.

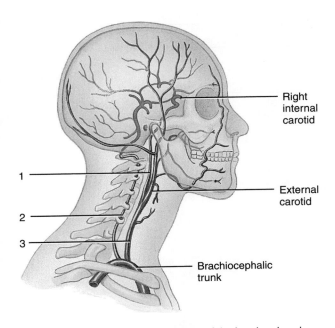

(a) Right lateral view of major arteries of the head and neck

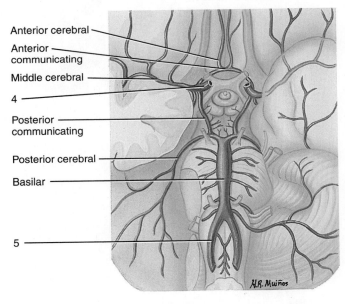

(b) Inferior view of brain showing cerebral arterial circle

(a) Lateral view
• common carotid
• internal carotid
• vertebral

1 _____

2 _____

3 _____

(b) Inferior view
• internal carotid
• vertebral

4 _____

5 _____

FIGURE 23.5   Arteries of the head and neck.

3. Major Arteries of the Upper Extremities

The subclavian artery is renamed the **axillary artery** within the axilla and continues as the **brachial artery** as it enters the arm. As the brachial artery enters the forearm, it divides to form the **radial artery** and the **ulnar artery.** The **superficial palmar arch** and the **deep palmar arch,** both of which supply blood to the fingers and palms, connect the radial and ulnar arteries at their distal ends and are examples of arterial anastomoses.

ACTIVITY 4 **ARTERIES OF THE UPPER EXTREMITIES**

1 Label the arteries on Figure 23.6.

2 Identify the arteries on models or charts.

3 Pronounce the names as you point to them.

4 Palpate your radial artery by placing the 2nd and 3rd fingers over the distal end of the radius.

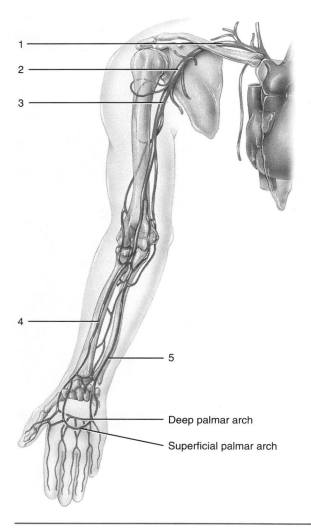

- axillary
- brachial
- radial
- subclavian
- ulnar

1 _____

2 _____

3 _____

4 _____

5 _____

Deep palmar arch

Superficial palmar arch

FIGURE 23.6 **Arteries supplying the upper extremities.**

4. Major Arterial Branches of the Descending Aorta

The diaphragm divides the aorta into thoracic and abdominal sections. The **thoracic aorta** lies to the left of the midline, just anterior to the vertebral column, and has branches that supply thoracic structures. The abdominal aorta is also anterior to the vertebral column but closer to the midline. It supplies the abdomen, pelvis, and lower extremities.

The major branches of the descending aorta are the **celiac trunk** (divides into the left gastric artery, splenic artery, and common hepatic artery), the **superior mesen-** teric artery, the paired **renal arteries,** and the **inferior mesenteric artery.** Within the pelvis, the abdominal aorta terminates as the paired **common iliac arteries.**

ACTIVITY 5 ARTERIAL BRANCHES OF THE DESCENDING AORTA

1 Label the arteries on Figure 23.7.

2 Identify the arteries on models or charts.

3 Pronounce the names as you point to them.

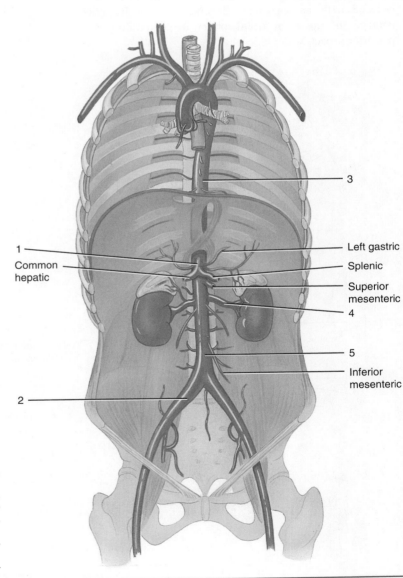

- abdominal aorta
- celiac trunk
- common iliac
- renal
- thoracic aorta

1 _____

2 _____

3 _____

4 _____

5 _____

FIGURE 23.7 **Descending aorta and major arterial branches.**

5. Major Arteries of the Pelvis and Lower Extremities

The **common iliac arteries** divide into the internal iliac and external iliac arteries. The **internal iliac arteries** are medial terminal branches off the common iliac arteries just anterior to the lumbosacral joint. They course posteriorly to the pelvis. The **external iliac artery,** the larger of the two arteries, becomes the femoral artery as it enters the thigh. The **femoral artery** descends along the middle of the anterior two thirds of the thigh and travels to the posterior thigh, becoming the popliteal artery as it enters the posterior knee area. The **popliteal artery** divides into the **anterior** and **posterior tibial arteries.** The anterior tibial artery travels to the anterior surface of the leg and descends to the ankles where it becomes the **dorsalis pedis artery.** The **posterior tibial artery** descends along the posterior aspect of the leg.

ACTIVITY 6 MAJOR ARTERIES OF THE PELVIS AND LOWER EXTREMITIES

1 Label the arteries on Figure 23.8.

2 Identify the arteries on models or charts.

3 Pronounce the names as you point to them.

4 Palpate the femoral artery pulse by placing your fingertips over the inguinal area.

5 Palpate the popliteal artery pulse by placing your fingertips on the posteror surface of the knee.

6 Palpate the pulse of the dorsal artery of the foot (dorsalis pedis artery) by placing your fingertips over the dorsal surface of the medial side of the foot near the ankle.

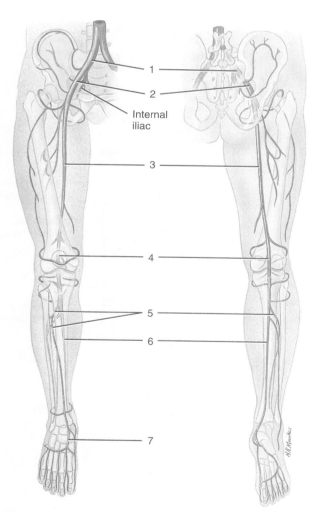

(a) Anterior view   (b) Posterior view

- anterior tibial
- common iliac
- dorsal artery of the foot (dorsalis pedis)
- external iliac
- femoral
- popliteal
- posterior tibial

1 _____

2 _____

3 _____

4 _____

5 _____

6 _____

7 _____

FIGURE 23.8   **Major arteries of the pelvis and lower extremities.**

C. SYSTEMIC VEINS

Veins carry blood from the capillaries back toward the heart. Veins can be found under the skin (superficial veins) or deep within the body, typically near an artery. Veins that accompany arteries usually have the same name as the artery. Veins carrying blood from the lungs to the left side of the heart carry oxygen-rich blood and are part of the pulmonary circulation, whereas veins carrying blood from all other body tissues carry oxygen-poor blood and are part of the systemic circulation.

> **CLINICAL NOTE:** *Thrombus in a deep vein can cause a pulmonary embolus if not treated with anticoagulant, whereas a thrombus in a superficial vein does not tend to form an embolus.*

1. Veins Carrying Blood to the Right Atrium of the Heart

Three large veins carry oxygen-poor blood into the right atrium—the **superior vena cava,** the **inferior vena cava,** and the **coronary sinus.**

ACTIVITY 7 VEINS CARRYING BLOOD TO THE HEART

1 Label the veins on Figure 23.9.

2 Identify the veins on models or charts.

3 Pronounce the names as you point to them.

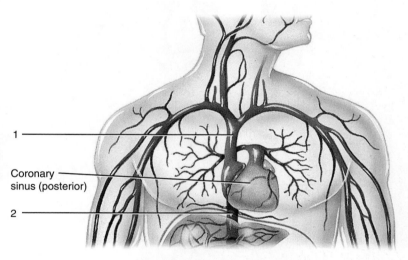

1 _____

Coronary sinus (posterior) _____

2 _____

- inferior vena cava
- superior vena cava

1 _____

2 _____

FIGURE 23.9   Systemic veins carrying blood to the heart.

2. Major Veins Draining the Head and Neck

Three large paired veins—the internal jugular veins, the external jugular veins, and the vertebral veins—drain blood from the head and neck. The **internal jugular veins** run lateral to the internal carotid and common carotid arteries, the **external jugular veins** are superficial veins descending along the lateral surface of the neck, and the **vertebral veins** descend through the transverse foramina of the vertebral column with the vertebral arteries.

ACTIVITY 8 MAJOR VEINS DRAINING THE HEAD AND NECK

1 Label the veins on Figure 23.10.

2 Identify the veins on models or charts.

3 Pronounce the names as you point to them.

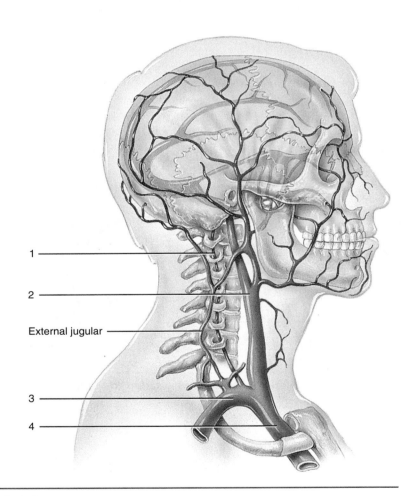

- brachiocephalic
- internal jugular
- subclavian
- vertebral

1 _____

2 _____

3 _____

4 _____

FIGURE 23.10 **Major veins of the head and neck.**

3. Major Veins Draining the Upper Extremities

Superficial veins are visible running beneath the skin of the arm. As we age, these veins become more visible. The major superficial veins of the upper extremities are the basilic and the cephalic veins. Each arm has a **cephalic vein** that travels along the lateral surface of the anterior limb and merges with the axillary vein inferior to the clavicle. Each **basilic vein** travels along the medial surface of the posterior forearm and the medial surface of the anterior arm. Each **median cubital vein,** a common site for obtaining blood samples, is anterior to the elbow and connects the basilic and cephalic veins.

The *major deep veins* of the arm and forearm run along with arteries and are named for the arteries they accompany. The **radial** and **ulnar veins,** found in each forearm, merge to form the **brachial veins.** The brachial veins ascend the arm and join with the basilic vein to form the **axillary vein.** The axillary vein becomes the **subclavian vein** as it leaves the axillary region.

ACTIVITY 9 MAJOR VEINS DRAINING THE UPPER EXTREMITIES

1 Label the veins on Figure 23.11.

2 Identify the veins on models or charts.

3 Pronounce the names as you point to them.

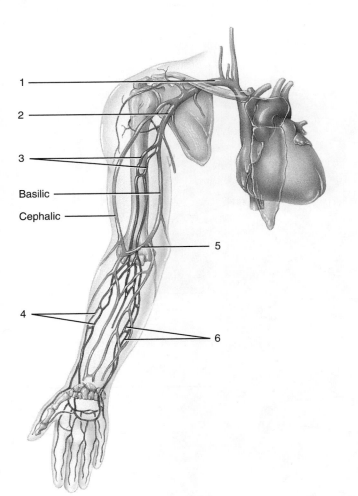

1

2

3

Basilic

Cephalic

5

4

6

- axillary
- brachial
- median cubital
- radial
- subclavian
- ulnar

1 _____

2 _____

3 _____

4 _____

5 _____

6 _____

FIGURE 23.11   **Major veins of the right upper extremity.**

4. Major Veins Draining the Thorax, Abdomen, and Pelvis

The **internal** and **external iliac veins** in the pelvis merge to form the **common iliac veins** that unite to form the **inferior vena cava.** The **renal veins** and **hepatic veins** are the major veins that drain directly into the inferior vena cava.

The veins draining the stomach, intestines, spleen, pancreas, and gallbladder do not drain directly into the inferior vena cava, but enter the **hepatic portal circulation** first. The major veins of the hepatic circulation are the inferior mesenteric vein, the splenic vein, the superior mesenteric vein, and hepatic portal vein. The **inferior mesenteric vein** drains blood from the large intestine and joins the **splenic vein** that carries blood from the stomach, pancreas, and spleen.

The **superior mesenteric vein** drains blood from the small intestine and merges with the splenic vein to form the **hepatic portal vein** that carries nutrient-rich blood to the liver for processing. The liver is drained by the **hepatic veins** that empty into the inferior vena cava.

The azygous system of veins includes the **azygous vein.** The azygous system of veins drains thoracic structures.

ACTIVITY 10 MAJOR VEINS DRAINING THE THORAX, ABDOMEN, AND PELVIS

1 Label the veins on Figure 23.12.

2 Label the hepatic portal system veins on Figure 23.13.

3 Identify the veins on models or charts.

4 Pronounce the names as you point to them.

• inferior vena cava
• left brachiocephalic
• left common iliac
• left external iliac
• left renal
• right brachiocephalic
• superior vena cava

1 _____

2 _____

3 _____

4 _____

5 _____

6 _____

7 _____

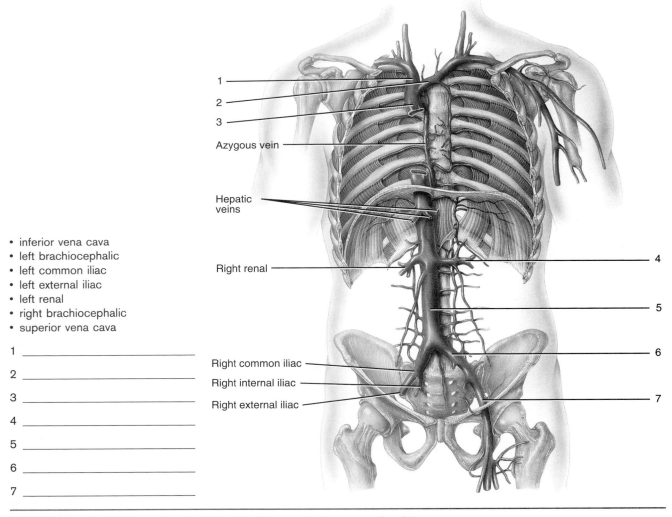

FIGURE 23.12   Major veins draining the thorax, abdomen, and pelvis.

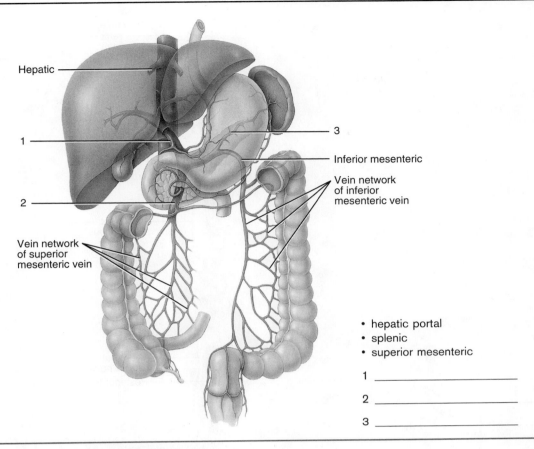

Hepatic

1

2

3

Inferior mesenteric

Vein network
of inferior
mesenteric vein

Vein network
of superior
mesenteric vein

• hepatic portal
• splenic
• superior mesenteric

1 _____

2 _____

3 _____

FIGURE 23.13   Hepatic portal circulation.

5. Major Veins Draining the Lower Extremities

The **great saphenous vein,** the major *superficial vein* of the leg, is the longest vein in the body and travels along the medial surface of the leg and the thigh. The *deep veins* of the leg ascend adjacent to the arteries of the same name. The **anterior** and paired **posterior tibial veins** ascend in the anterior and posterior leg, and unite inferior to the popliteal fossa to form the popliteal vein. The **popliteal vein** ascends along the posterior surface of the knee and becomes the **femoral vein** that travels up the posterior thigh and becomes the **external iliac vein** in the pelvis.

ACTIVITY 11 MAJOR VEINS DRAINING THE LOWER EXTREMITIES

1 Label the veins on Figure 23.14.

2 Identify the veins on models or charts.

3 Pronounce the names as you point to them.

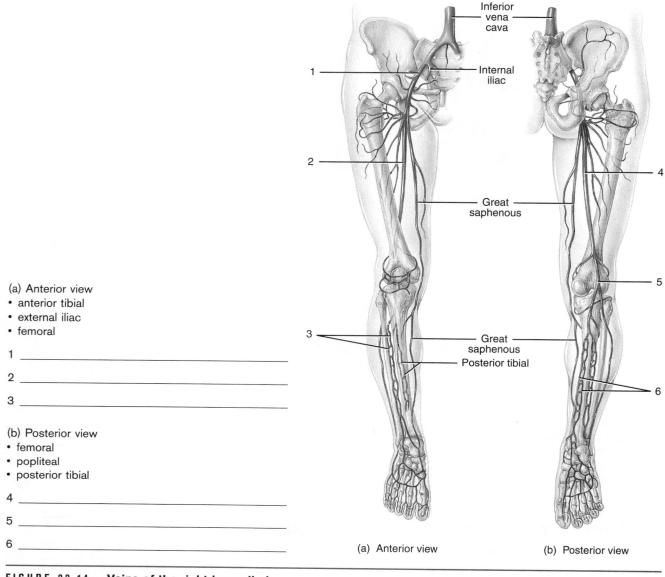

(a) Anterior view
• anterior tibial
• external iliac
• femoral

1 _____

2 _____

3 _____

(b) Posterior view
• femoral
• popliteal
• posterior tibial

4 _____

5 _____

6 _____

(a) Anterior view   (b) Posterior view

FIGURE 23.14   Veins of the right lower limb.

D. PULMONARY CIRCULATION

The pulmonary circulation carries oxygen-poor blood from the right ventricle to the capillaries of the lung where oxygen is added and carbon dioxide is removed. The **pulmonary trunk** carries oxygen-poor blood from the right ventricle and divides to form the **right** and **left pulmonary arteries** that carry blood to the lungs. Oxygen-rich blood leaves the lungs through the **pulmonary veins** that drain into the left atrium.

ACTIVITY 12 MAJOR BLOOD VESSELS OF THE PULMONARY CIRCULATION

1 Label the blood vessels on Figure 23.15.

2 Identify the blood vessels on models or charts.

3 Pronounce the names as you point to them.

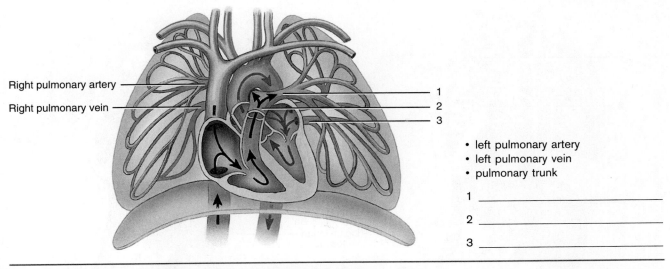

Right pulmonary artery

Right pulmonary vein

1
2
3

- left pulmonary artery
- left pulmonary vein
- pulmonary trunk

1 _____

2 _____

3 _____

FIGURE 23.15   Blood vessels of the pulmonary circulation.

E. FETAL CIRCULATION

Because the fetus cannot breathe or ingest food, the mother's circulatory system provides the fetus with oxygen and nutrients and eliminates carbon dioxide and other wastes from fetal blood. This exchange occurs across the **placenta** that forms within the uterus during early pregnancy. Uterine blood vessels enter the placenta, and substances are exchanged by diffusion between maternal and placental capillaries without direct mixing of maternal and fetal blood. Two **umbilical arteries** carry oxygen-poor fetal blood to the placenta, and one **umbilical vein** carries the oxygen-rich blood back to the fetus. The **ductus venosus** allows most of the blood to bypass the liver and enter the inferior vena cava. The **foramen ovale** and the **ductus arteriosus** allow blood to bypass the fetal lungs, and after birth, vascular changes occur to allow blood to enter the lungs.

ACTIVITY 13 FETAL CIRCULATION

1 Label the structures on Figure 23.16.

2 Identify the structures on models or charts.

3 Pronounce the names as you point to them.

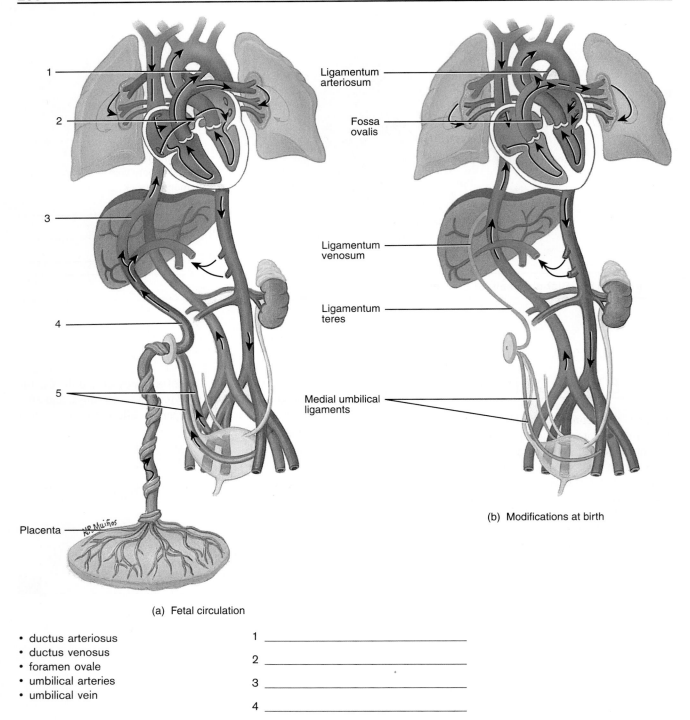

Ligamentum arteriosum

Fossa ovalis

Ligamentum venosum

Ligamentum teres

Medial umbilical ligaments

(b) Modifications at birth

Placenta

(a) Fetal circulation

- ductus arteriosus
- ductus venosus
- foramen ovale
- umbilical arteries
- umbilical vein

1 _____

2 _____

3 _____

4 _____

5 _____

FIGURE 23.16   Fetal circulation and modifications after birth.

REVIEWING YOUR KNOWLEDGE

A. Structure of Arteries, Veins, and Capillaries

Mark the following True or False. If False, underline and change word to make statement True.

_____ **1.** Venous valves prevent backflow of blood.

_____ **2.** Walls of veins are thicker than walls of arteries of the same size.

_____ **3.** Lumens of veins are larger than lumens of arteries of the same size.

_____ **4.** Capillary walls are composed of an endothelium and a basement membrane only.

_____ **5.** Lumen size increases during vasoconstriction.

B. Artery Identification

Label the arteries in Figure 23.17.

1. _____

2. _____

3. _____

4. _____

5. _____

6. _____

7. _____

8. _____

9. _____

10. _____

11. _____

12. _____

13. _____

14. _____

15. _____

16. _____

17. _____

18. _____

19. _____

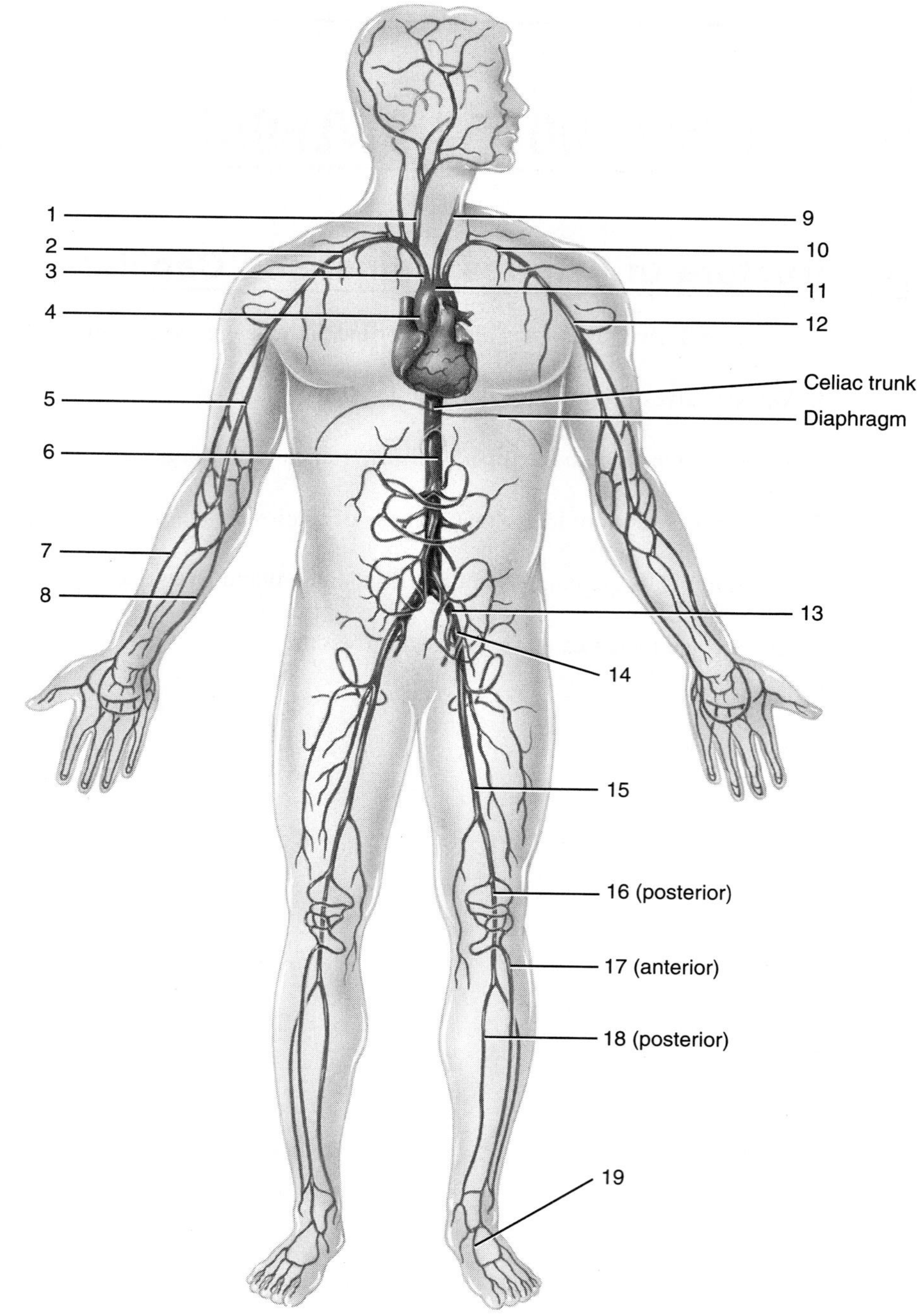

FIGURE 23.17   **Major systemic arteries.**

C. Vein Identification

Label the veins in Figure 23.18.

1. _____

2. _____

3. _____

4. _____

5. _____

6. _____

7. _____

8. _____

9. _____

10. _____

11. _____

12. _____

13. _____

14. _____

15. _____

16. _____

17. _____

18. _____

19. _____

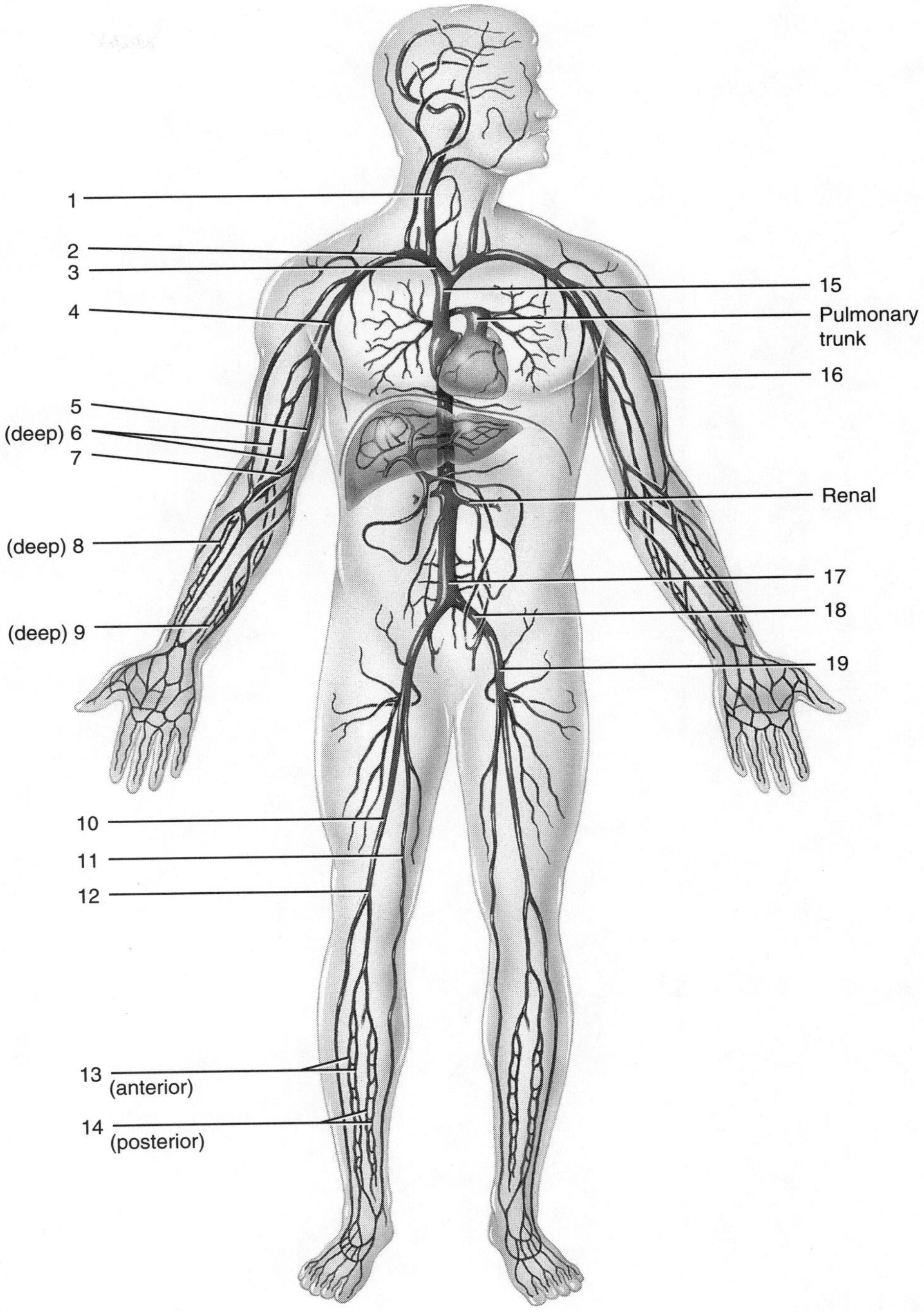

FIGURE 23.18   Major systemic veins.

STRUCTURE OF THE RESPIRATORY SYSTEM

EXERCISE

OBJECTIVES

After completing this exercise, you should be able to:

- Locate and identify the organs of the respiratory system on models or charts
- Describe the structure and function of the respiratory system organs
- Trace the path of air from the nose to the alveoli

MATERIALS

- models or charts with respiratory organs
- sagittal section model or chart of the human head, neck, and thorax; larynx model
- sheep or pig pluck
- colored pencils

A. GROSS ANATOMY OF THE RESPIRATORY ORGANS

The respiratory system moves air into and out of the body. It is also the site where oxygen diffuses into the bloodstream, and carbon dioxide diffuses out of the bloodstream to be exhaled.

1. Nose

Nasal bones form the base of the **nose,** and hyaline cartilage forms the anterior portion. Two **external nares** (singular, naris) are the openings for air to enter a space called the **nasal cavity** that is lined with a mucous membrane. The **nasal septum** separates this cavity into right and left portions. On either side of the nose are the **nasal conchae** (*concha* = shell-shaped), or **turbinates** (*turbinum* = scroll-shaped), three curved bony structures that extend into the nasal cavity. Beneath each nasal concha is a concavity called a **meatus** (*meatus* = passage) that increases surface area in the nose. The bony **hard palate** forms the floor of the nasal cavity. At the junction of the hard and soft palates are the **internal nares,** two openings which lead into the posterior nasal pharynx. The frontal, maxillary, ethmoidal, and sphenoidal bones have mucous membrane-lined **paranasal sinuses** that have ducts that drain into the nasal cavity.

2. Pharynx

The **pharynx,** or throat, is divided into three regions: the nasopharynx (*naso-* = nose), oropharynx (*oro-* = mouth), and laryngopharynx (*laryngo-* = larynx), corresponding to the anatomical structures nearby. The **nasopharynx** begins at the internal nares and ends at the soft palate. The **oropharynx** begins at the soft palate and extends to the level of the hyoid bone, and the **laryngopharynx** extends from the hyoid bone to the beginning of the esophagus.

Several important structures are found within or open into the pharynx. Two **auditory tubes** (Eustachian tubes) connect the middle ear and the nasopharynx, and a single **pharyngeal tonsil** (adenoid) is located in the posterior wall of the nasopharynx. The oropharynx has two pairs of tonsils, the **palatine** (palate) and **lingual** (tongue) tonsils. The muscular **soft palate** is found in the oropharynx posterior to the bony hard palate. The oval-shaped **uvula** dangles inferiorly as an extension of the soft palate. During swallowing, the soft palate pushes superiorly to close off the nasopharynx and to direct food toward the laryngopharynx. Note that the mucous membranes of the nose, paranasal sinuses, pharynx, and middle ear are connected by ducts, which explains why infections can spread to any adjacent areas.

3. Larynx

The **larynx,** or voice box, connects the laryngopharynx with the trachea. This organ consists of nine hyaline cartilages and houses the vocal cords. The larynx contains three small, paired cartilages located in the posterior wall of the larynx and three single cartilages—the thyroid cartilage, cricoid cartilage, and epiglottis—which comprise the main body of the larynx. The largest cartilage, the **thyroid cartilage** (*thyroid* = shield-shaped), is seen anteriorly and has a prominence called the Adam's apple that is made of hyaline cartilage. The **cricoid cartilage** (*cricoid* = ring-like), which is inferior to the thyroid cartilage, is larger on the posterior side than on the anterior side and consists of hyaline cartilage. As the term "cricoid" suggests, this is the only cartilage that is a complete ring in the larynx or trachea. The oval-shaped **epiglottis** (*epi-* = over; *-glottis* = tongue) is a flap-like structure composed of elastic cartilage and closes over the glottis during swallowing. The **glottis** includes the **vocal folds** or **true vocal cords** and the slit-like opening between them that allows air into the trachea.

ACTIVITY 1 NOSE, PHARYNX, AND LARYNX

1 Label the three divisions of the pharynx in Figure 24.1(a).

2 Label the structures of the nose and pharynx in Figure 24.1(b).

3 Using constrasting colors, color the structures in Figures 21.1(a) and (b).

4 Label the structures of the larynx in Figures 24.2(a) and (b).

5 Identify the structures on a sagittal section model or anatomical chart of the human head and larynx model.

6 Pronounce the terms as you point to them on the model.

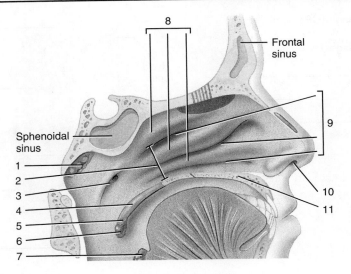

(a) Sagittal section

- external naris (NEH-ris)
- hard palate
- nasal conchae or turbinates (CON-chee; TUR-bin-its)
- nasal meatuses (mee-AY-tuses)
- internal naris
- lingual tonsil
- palatine tonsil
- pharyngeal tonsil (fair-IN-gee-al)
- opening of auditory tube
- soft palate
- uvula (YOU-view-la)

1 _____

2 _____

3 _____

4 _____

5 _____

6 _____

7 _____

8 _____

9 _____

10 _____

11 _____

- laryngopharynx (la-rin-go-FAIR-inks)
- nasopharynx
- oropharynx (oro-FAIR-inks)

12 _____

13 _____

14 _____

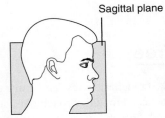

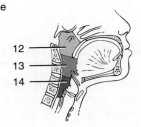

(b) Divisions of the pharynx

FIGURE 24.1   Nose and pharyngeal structures.

- cricoid cartilage (CRY-koid)
- epiglottis
- thyroid cartilage

1 _____

2 _____

3 _____

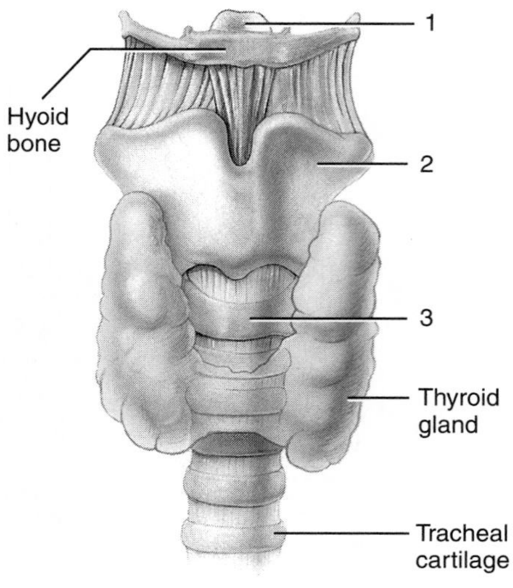

(a) Anterior

- glottis
- vocal folds

4 _____

5 _____

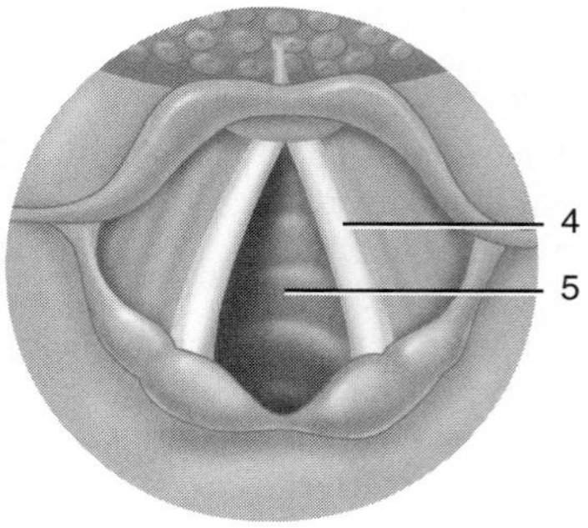

(b) Superior

FIGURE 24.2 The larynx and associated structures.

4. Bronchial Tree

Just as a tree has a trunk, branches, twigs, and fruit, the lower airways have a trachea, bronchi, bronchioles, and alveoli. If the trachea and bronchi were inverted, they would resemble the branches of a tree, hence the name **bronchial tree.** The **trachea** (*trachea* = sturdy), or windpipe, is a tube-like conduit that conducts air from the larynx to the bronchi. It is located anterior to the esophagus and can be palpated with your hand on the surface of your neck. When palpating the trachea you will feel bumps, the C-shaped **tracheal cartilages** that keep the airway open. The trachea ends and branches into a shorter right and longer left primary bronchus, each serving the correspon-

ding lung. Each **primary bronchus** branches into **secondary (lobar) bronchi** to supply a **lobe** of each lung. The right primary bronchus branches into three secondary bronchi, and the left primary bronchus branches into two secondary bronchi. The secondary bronchi further subdivide into **tertiary (segmental) bronchi** that divide into **bronchioles** each serving small compartments of the lung called **lobules.** Bronchioles further subdivide into **terminal bronchioles** which branch into **respiratory bronchioles** that begin the respiratory zone of the lung. Respiratory bronchioles further divide into **alveolar ducts** that lead into clusters of alveoli called **alveolar sacs. Alveoli** bud off alveolar sacs like individual grapes in a grape cluster.

5. Lungs

The two **lungs** are divided into *three lobes* on the right side and *two lobes* on the left. The three right lobes are the **superior, middle,** and **inferior lobes,** and the left lobes are called the **superior** and **inferior lobes.** The rounded superior part of the lung is the **apex** and the broader inferior part is the **base** that rests on the diaphragm. Each lung has a **hilus,** an area surrounded with pleura, where the bronchi,

blood and lymphatic vessels, and nerves enter or exit the medial side of the lung.

The **parietal pleura** lines the thoracic cavity wall, and **visceral pleura** covers the surface of each lung. The **pleural cavity** is the space between the two pleural layers that contains pleural fluid.

- apex
- base
- diaphragm
- inferior lobe
- larynx
- middle lobe
- parietal pleura
- pleural cavity
- primary bronchus
- secondary bronchus
- superior lobe
- tertiary bronchus
- trachea
- visceral pleura

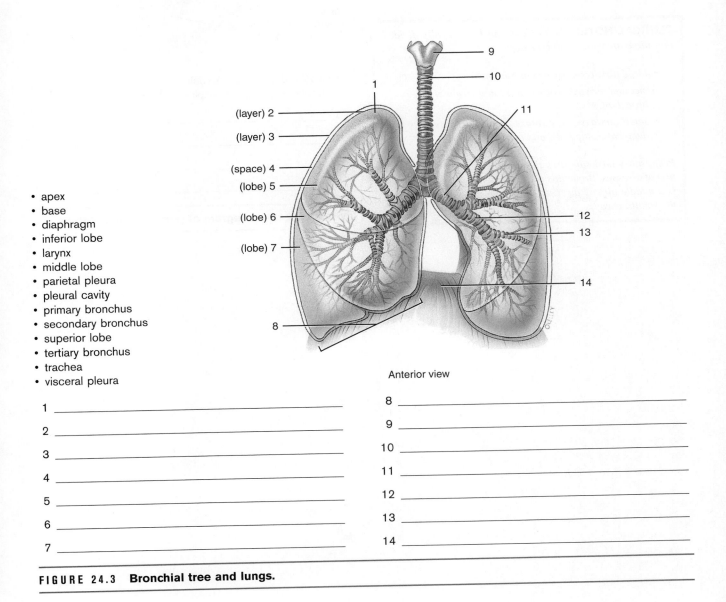

Anterior view

1 _____
2 _____
3 _____
4 _____
5 _____
6 _____
7 _____

8 _____
9 _____
10 _____
11 _____
12 _____
13 _____
14 _____

FIGURE 24.3   Bronchial tree and lungs.

ACTIVITY 2 BRONCHIAL TREE AND LUNGS

1 Label the respiratory structures in Figures 24.3 and 24.4.

2 Using contrasting colors, color the structures in Figures 24.3 and 24.4.

3 Locate the structures on lung models or anatomical charts.

4 Pronounce each term as you point to it.

5 Observe the sheep or pig pluck demonstration with the bronchial tree dissected.

CLINICAL NOTE: *The airways can be clinically divided into three areas according to size:*

- *large airways—trachea and bronchi*
- *medium airways—bronchioles and terminal bronchioles*
- *small airways—respiratory bronchioles, alveolar ducts, alveolar sacs, and alveoli*

Respiratory medicine uses these terms to identify the location of diseases. Bronchitis is a large airway disease, asthma is a middle airway disease, and emphysema is a disease of the small airways.

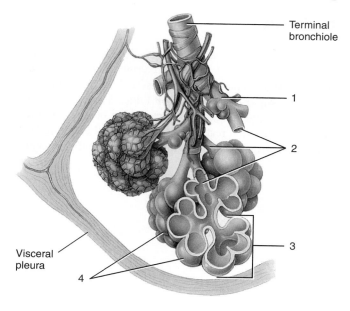

- alveolar ducts
- alveolar sac
- alveoli
- respiratory bronchiole

1 _____

2 _____

3 _____

4 _____

FIGURE 24.4   Diagram of portion of lung lobule.

REVIEWING YOUR KNOWLEDGE

A. Air Flow

Trace the air flow through the respiratory system starting with the external nares. Number the structures 1 through 17.

_____ alveolar duct

_____ alveolar sac

_____ alveolus

_____ bronchiole

_____ external nares

_____ internal nares

_____ larynx

_____ laryngopharynx

_____ nasal cavity

_____ nasopharynx

_____ oropharynx

_____ primary bronchus

_____ respiratory bronchiole

_____ secondary bronchus

_____ terminal bronchiole

_____ tertiary bronchus

_____ trachea

B. Structure and Function

Write the name of the structure that the phrase describes.

alveoli
bronchiole
epiglottis
larynx
tracheal cartilages
oropharynx
tertiary bronchi
trachea

_____ **1.** Connects the laryngopharynx with the trachea

_____ **2.** Tube-like structure that conducts air from the larynx to the bronchi

_____ **3.** Closes over the glottis during swallowing

_____ **4.** Keep the trachea from collapsing

_____ **5.** Division of the bronchi that subdivides into bronchioles

_____ **6.** Conducts air from the nasopharynx to the laryngopharynx

_____ **7.** Small, round sacs where gas exchange occurs

_____ **8.** Small airway that serves a lobule

C. Application

1. Explain why food or liquid in the mouth can be expelled from the nostrils if the mouth is closed when the person laughs.

2. Explain how an infection in the nasopharynx can also result in an infection in the paranasal sinuses and/or the middle ear.

PULMONARY VENTILATION

25

OBJECTIVES

After completing this exercise, you should be able to:

- Explain how changes in thoracic and lung volumes and lung pressures result in pulmonary ventilation
- Define and measure or calculate lung volumes and capacities

MATERIALS

- models or charts with respiratory muscles
- bell jar model of lungs
- handheld dry spirometer, disposable mouthpieces, 70% alcohol and cotton, or alcohol wipes
- colored pencils

The respiratory system supplies oxygen needed by body cells to produce adenosine triphosphate (ATP) for metabolism and removes carbon dioxide produced by metabolic reactions. Respiration involves three steps—pulmonary ventilation, external respiration, and internal respiration—and requires the cardiovascular system to transport oxygen and carbon dioxide throughout the body. **Pulmonary ventilation,** or breathing, is the movement of air between the atmosphere and the lungs that occurs when we inhale (inspiration) and exhale (expiration).

A. PULMONARY VENTILATION

During pulmonary ventilation, air moves from an area of higher pressure to an area of lower pressure. Changes in air pressure in the lung (alveolar pressure) occur when lung volume changes. When lung volume increases, the pressure of the air inside decreases and air flows into the lungs. When lung volume decreases, the pressure of the air inside increases and air flows out of the lungs.

1. Changing Thoracic and Lung Volumes

When the diaphragm and other respiratory muscles contract or relax, they change the size of the thorax, which in turn changes lung volume. **Normal inhalation** is caused mainly by contraction of the diaphragm. The **diaphragm** is dome-shaped when relaxed and flattens when contracted. When the diaphragm flattens, the length of the thoracic cavity and its volume increase and air flows into the lungs. During **forced inhalation,** contraction of the **external intercostals** further increases the width of the thoracic cage by raising the ribs further increasing lung volume and resulting in a greater volume of inhaled air.

Normal exhalation is a passive process involving relaxation of the diaphragm and elastic recoil of the chest wall and lungs. The diaphragm becomes dome-shaped and decreases the length of the thoracic cavity and thoracic volume, causing air to flow out of the lungs. In **forced exha-** lation, contraction of the **internal intercostals** depresses the rib cage. In addition, contraction of the **abdominal muscles (external oblique, internal oblique, transverse abdominis,** and **rectus abdominis)** push the diaphragm superiorly, further decreasing thoracic volume and resulting in a greater volume of exhaled air.

ACTIVITY 1 RESPIRATORY MUSCLES AND VOLUME CHANGES

1 Label the respiratory muscles in Figure 25.1.

2 Using contrasting colors, color the respiratory muscles in Figure 25.1.

3 Identify the respiratory muscles on models or charts.

4 Pronounce the muscle names as you point to them.

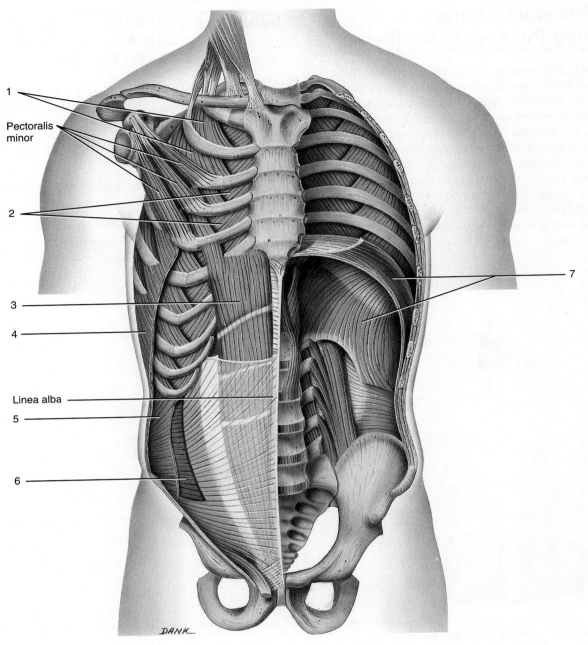

Pectoralis minor

Linea alba

(a) Anterior superficial view   (b) Anterior deep view

- diaphragm
- external intercostals
- external oblique
- internal intercostals
- internal oblique
- rectus abdominis
- transverse abdominis

1 _____   5 _____

2 _____   6 _____

3 _____   7 _____

4 _____

FIGURE 25.1   Respiratory muscles.

2. Pressure Changes During Pulmonary Ventilation

As the thorax expands, the parietal pleura attached to the internal thoracic wall is pulled outward. The **pleural cavity** increases slightly in volume causing a decrease in **intrapleural pressure,** the pressure between the pleural layers. The decrease in intrapleural pressure and the surface tension of pleural fluid cause the visceral pleura, and therefore the lungs, to also be pulled outward. As the lungs increase in volume, the **alveolar (intrapulmonic) pressure** decreases to below atmospheric pressure, and air enters the lungs. When the thorax decreases in size, the lungs recoil, causing intrapleural pressure and alveolar pressure to increase, and air leaves the lungs.

ACTIVITY 2 PRESSURE CHANGES DURING PULMONARY VENTILATION

1 Using the model lung (bell-jar model), observe what happens when the rubber diaphragm at the base of the model is domed or flattened.

2 Label the model lung in Figure 25.2 with the analogous respiratory system structures.

3 Complete Table 25.1 by circling the correct choice in the column "Movement of Diaphragm."

4 Answer discussion questions concerning model lung demonstration with your lab partners.

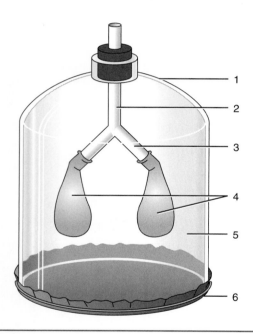

- bronchus
- diaphragm
- pleural cavity
- lungs covered with visceral pleura
- parietal pleura and thoracic wall
- trachea

1 _____

2 _____

3 _____

4 _____

5 _____

6 _____

FIGURE 25.2   Model lung (bell-jar model).

TABLE 25.1 Bell Jar (Model Lung) Demonstration

OBSERVATIONS	MOVEMENT OF DIAPHRAGM	
	AS DIAPHRAGM DOMES (*circle correct choice*)	**AS DIAPHRAGM FLATTENS** (*circle correct choice*)
Volume of air in bell jar	1. Increased or decreased	5. Increased or decreased
Pressure of air in bell jar	2. Increased or decreased	6. Increased or decreased
Balloon size	3. Inflated or deflated	7. Inflated or deflated
Direction of air flow	4. Into or out of balloons	8. Into or out of balloons

❓ DISCUSSION QUESTIONS: BELL JAR (MODEL LUNG) DEMONSTRATION

1 (a) In the model lung demonstration, the glass jar is filled with air. What is in the pleural cavity in the thorax? (b) Compare the relative size of the pleural space in the model lung to that in the human lung.

2 (a) In the model lung, is the air in the glass jar at atmospheric pressure or sub-atmospheric pressure? (b) Is the pressure in the pleural cavity atmospheric or sub-atmospheric?

3 (a) In humans, is there one pleural cavity surrounding both lungs or two separate pleural cavities, one surrounding each lung? (b) Does this differ from the model lung?

4 In the model lung demonstration, the glass bell jar does not change in size. Describe the changes in the corresponding human respiratory system structures during inhalation and exhalation.

B. MEASURING LUNG VOLUMES WITH SPIROMETRY

Lung volumes and capacities are defined in Table 25.2. Capacities are combinations of different lung volumes. These volume measurements vary according to gender, age, height, and physical condition.

TABLE 25.2 Lung Volumes and Capacities

VOLUME OR CAPACITY	ABBREVIATION	DEFINITION	AVERAGE VALUES FOR HEALTHY ADULT
Tidal volume	TV	The amount of air inhaled and exhaled during one normal breath at rest.	500 mL
Inspiratory reserve volume	IRV	Maximum amount of air that can be inhaled after a normal inhalation.	3,100 mL
Expiratory reserve volume	ERV	Maximum amount of air that can be exhaled after a normal exhalation.	1,200 mL
Residual volume	RV	Amount of air that remains in the lungs after a maximal exhalation.	1,200 mL
Inspiratory capacity	IC	IC = TV + IRV	3,600 mL
Functional residual capacity	FRC	FRC = RV + ERV	2,400 mL
Vital capacity	VC	VC = IRV + TV + ERV	4,800 mL
Total lung capacity	TLC	TLC = IRV + TV + ERV + RV	6,000 mL

1. Measuring Tidal Volume, Expiratory Reserve Volume, and Vital Capacity

A **spirometer** is an instrument used to measure lung volumes and capacities. There are different types of spirometers that vary on how they measure air volumes and capacities. In this activity, you will be using a dry, handheld spirometer. **Exhale only** into this type of spirometer. It does not measure inspiratory volumes. Spirometers that measure inspiratory air volumes must be cleaned or supplied with new filters when used by a different person to prevent spread of infection. Your instructor will supply you with instructions if you are using a different type of spirometer than the dry, handheld spirometer.

ACTIVITY 3 MEASURING LUNG VOLUMES AND CAPACITIES

1 Preparations for Activity 3

Prepare a dry, handheld spirometer:

- Wipe the nozzle of the spirometer with 70% alcohol.
- Place a clean, disposable mouthpiece on the nozzle.
- Use a noseclip or your fingers to close your nostrils to prevent air leaking out of your nose.
- When blowing into the spirometer, hold it in a horizontal position with the face of the dial upward.

- With use, water will condense inside the spirometer dial. Follow instructions included with your spirometer for removing the condensation.

2 Measure tidal volume (TV)

- Set the dial on the spirometer to 0.
- Take two or three normal breaths then inhale normally, place the spirometer to your lips and exhale normally into the spirometer. Repeat this process two more times, and **do not reset** the dial before each measurement. Divide the volume by 3 to obtain the average TV. Record the average TV in Table 25.3.

3 Measure expiratory reserve volume (ERV)

- Reset the dial to 0.
- Take two or three normal breaths, then inhale and exhale normally. Without taking another breath, place the spirometer to your lips and exhale as much air as you can. Record value in Table 25.3. Repeat this process two more times, resetting the dial to 0 before each measurement. Calculate and record the average ERV.

4 Measure vital capacity (VC)

- Reset the dial to 0.
- Take two or three normal breaths, then inhale as much air as possible. Then put the spirometer to your lips and forcefully exhale as much air as you can as fast as you can. Record the value for VC in Table 25.3. Repeat this process two more times, resetting the dial to 0 before each measurement. Calculate and record the average VC.

TABLE 25.3 Measured Lung Volumes and Capacities

MEASUREMENT	VALUE 1	VALUE 2	VALUE 3	AVERAGE VALUE
TV				
ERV				
VC				

2. Calculating Inspiratory Reserve Capacity, Inspiratory Capacity, Functional Residual Capacity, Total Lung Capacity

Dry handheld spirometers cannot be used to measure inspiratory volumes and capacities. However, inspiratory reserve capacity (IRV), inspiratory capacity (IC), functional residual capacity (FRC), and total lung capacity (TLC) can be calculated using measured values for TV, ERV, and VC and using 1,200 mL for residual volume (RV). As indicated in Table 25.2, 1,200 mL is the average RV for healthy adults.

ACTIVITY 4 CALCULATING LUNG VOLUMES AND CAPACITIES

Use your measured average values for TV, ERV, and VC for the calculations. Use 1,200 mL for the RV.

1 Calculate the IRV and record the value in Table 25.4. The equation for VC in Table 25.2 was used to formulate the equation for IRV.

$$IRV = VC - (TV + ERV).$$

2 Use the equations in Table 25.2 to calculate IC, FRC, and TLC. Record the values in Table 25.4.

TABLE 25.4 Calculated Lung Volumes and Capacities

VOLUME OR CAPACITY	EQUATION	VALUE (mL)
IRV		
IC		
RV		
FRC		
TLC		

REVIEWING YOUR KNOWLEDGE

A. Function of Respiratory Muscles

Write the name of the respiratory muscle that corresponds to the appropriate muscle function.

abdominal muscles
diaphragm
external intercostals
internal intercostals

_____ **1.** Main inspiratory muscle.

_____ **2.** Accessory muscle of inspiration.

_____ **3.**
        } Muscles used in forced expiration.
_____ **4.**

B. Respiratory Muscle Actions on Thoracic Size

Complete Table 25.5 by circling the correct choice in the column "Effect on Thoracic Dimensions."

TABLE 25.5 Respiratory Muscle Functions

MUSCLE	FUNCTION	EFFECT ON THORACIC DIMENSIONS (circle correct choice)
External intercostals	Elevate ribs	Increase or decrease diameter
Internal intercostals	Depress ribs	Increase or decrease diameter
Diaphragm	Flattens when contracted	Increase or decrease length
Abdominal muscles	Compress abdominal contents, increase abdominal pressure, and force diaphragm superiorly	Increase or decrease length

C. Volume and Pressure Changes During Pulmonary Ventilation

Indicate whether each volume or pressure in Table 25.6 increases or decreases during inhalation and exhalation.

TABLE 25.6 Volume and Pressure Changes During Pulmonary Ventilation		
VOLUME OR PRESSURE	INHALATION	EXHALATION
Thoracic volume	1.	2.
Intrapleural cavity volume	3.	4.
Intrapleural pressure	5.	6.
Lung volume	7.	8.
Alveolar (intrapulmonic) pressure	9.	10.

D. Lung Volumes and Capacities

Write the lung volume or capacity that corresponds with the definition.

expiratory reserve volume
functional residual capacity
inspiratory capacity
inspiratory reserve volume
residual volume
tidal volume
total lung capacity
vital capacity

_____ **1.** Sum of RV + ERV.

_____ **2.** Maximum amount of air that can be exhaled after a maximal inhalation.

_____ **3.** Sum of IRV + TV + ERV + RV.

_____ **4.** Amount of air that moved into the lungs during inhalation or out of lungs during exhalation.

_____ **5.** Sum of TV and IRV.

_____ **6.** Amount of air that remains in the lungs after a maximal exhalation.

_____ **7.** Maximum amount of air that can be inhaled after a normal inhalation.

_____ **8.** Maximum amount of air that can be exhaled after a normal exhalation.

STRUCTURE OF THE DIGESTIVE SYSTEM

26

OBJECTIVES

After completing this exercise, you should be able to:

- Identify the major gastrointestinal (GI) tract organs and accessory digestive organs on models or charts

- Describe the peritoneum and identify the major peritoneal structures

- Describe the functions of each GI tract organ and accessory digestive organ

MATERIALS

- human torso
- sagittal head section model
- colored pencils

A. OVERVIEW OF THE DIGESTIVE SYSTEM

The **digestive system** contains the organs of the gastrointestinal (GI) tract and the accessory digestive organs. The **GI tract,** or **alimentary** (*alimentary* = nourishment) **canal,** is a tube that extends from the mouth to the anus. The lumen of the GI tract opens to the external environment at either end, and anything inside the lumen is therefore considered to be external to the body. The **GI tract organs** include the mouth, most of the pharynx, esophagus, stomach, small intestine, large intestine, and anus. The **accessory digestive organs**—the salivary glands, tongue, teeth, pancreas, liver, and gallbladder—assist in digestion.

ACTIVITY 1 ORGANS OF THE DIGESTIVE SYSTEM

1 Label and review the structures in Figure 26.1.

2 Using contrasting colors, color the digestive organs in Figure 26.1.

3 Identify the structures on models or charts.

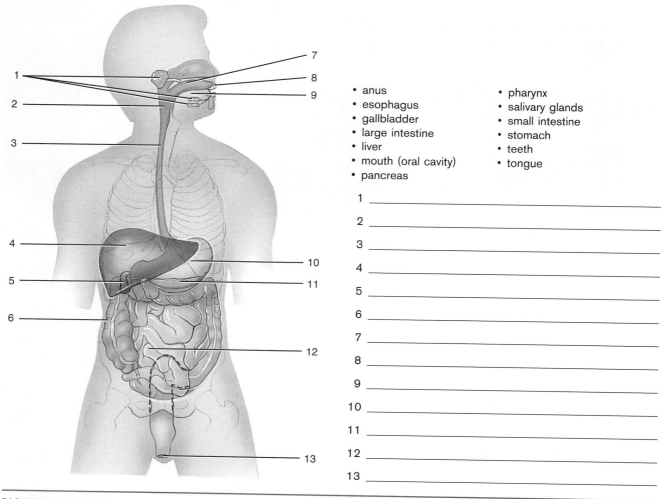

- anus
- esophagus
- gallbladder
- large intestine
- liver
- mouth (oral cavity)
- pancreas

- pharynx
- salivary glands
- small intestine
- stomach
- teeth
- tongue

1 _____

2 _____

3 _____

4 _____

5 _____

6 _____

7 _____

8 _____

9 _____

10 _____

11 _____

12 _____

13 _____

FIGURE 26.1   Organs of the digestive system.

B. THE PERITONEUM

The largest serous membrane in the body is the **peritoneum,** which is located in the abdominopelvic cavity. As with all serous membranes, it has a parietal and a visceral layer. The **parietal peritoneum** lines the inner surface of the abdominopelvic wall, and the **visceral peritoneum** (serosa) covers organs within the abdominopelvic cavity. **Peritoneal** (serous) **fluid** is found between the two peritoneal layers in the **peritoneal cavity.** Organs that lie outside the peritoneal cavity are called **retroperitoneal** (*retro-* = behind) **organs** (i.e., pancreas and kidneys).

The peritoneum has large folds not found in other serous membranes that secure organs together and to the abdominal walls. Two peritoneal folds are the falciform ligament and the greater omentum. The liver is bound to the anterior abdominal wall by the **falciform** (*falc-* = sickle-shaped) **ligament.** The **greater omentum** (*omentum* = fat skin) is

a large, fatty-looking apron that folds back on itself and attaches to the transverse colon. It forms a pouch that hangs between the body wall and the anterior surface of the small intestine. The **mesentery** is a part of the peritoneum that holds the small intestine to the posterior abdominal wall.

ACTIVITY 2 THE PERITONEUM

1 Label the structures on Figures 26.2(a) and (b).

2 Using contrasting colors, color the peritoneal folds in Figures 26.2(a) and (b).

3 Identify the structures on models or charts.

4 Pronounce the name of each term as you point to it.

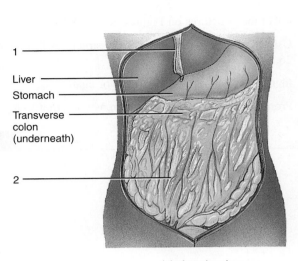

1

Liver

Stomach

Transverse
colon
(underneath)

2

(a) Anterior view

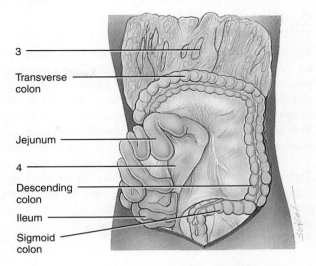

3

Transverse
colon

Jejunum

4

Descending
colon

Ileum

Sigmoid
colon

(b) Anterior view with greater omentum
lifted and small intestine reflected
to right side

- falciform ligament
- greater omentum
- greater omentum (reflected)
- mesentery

1 _____

2 _____

3 _____

4 _____

FIGURE 26.2 **Peritoneal folds.**

C. GASTROINTESTINAL TRACT ORGANS

1. Mouth

Food first enters the digestive system through the mouth. The **hard** and **soft palates** form the roof of the mouth, the tongue forms the floor, and the **cheeks** form the lateral walls of the mouth. The anterior border is the **lips,** and the posterior border is the **fauces** (*fauces* = passages), which is the opening to the oropharynx. The mouth is also called the **oral** or **buccal** (*bucca-* = cheeks) **cavity proper.** The palatine bone and the palatine processes of the maxillae form the **hard palate,** and the soft palate (smooth muscle) is posterior to this. The muscular **soft palate** forms the two arches that border the fauces, and also hangs down to form the oval process called the **uvula.** During swallowing, the soft palate keeps food from entering the nasal cavity. When you look into a mirror with your mouth wide open, you see the two arches, with the uvula hanging down in between. The **palatine tonsils** are located between the two arches.

2. Pharynx

The pharynx is divided into three specific areas, named after structures that are in close proximity. The **nasopharynx,** the superior area, is posterior and inferior to the nose and has respiratory functions. The middle area, the **oropharynx,** is posterior to the fauces of the oral cavity. This area has both respiratory and digestive functions because air, food, and liquid pass through it. The inferior area is the **laryngopharynx,** located posterior to the larynx. This area is also a common passageway for air, food, and liquid. During swallowing, the epiglottis of the larynx closes off the inferior portion of the laryngopharynx, allowing food and drink to enter the esophagus.

ACTIVITY 3 MOUTH AND PHARYNX

1 Label the structures on Figures 26.3 and 26.4.

2 Using contrasting colors, color the structures in Figure 26.3 and 26.4.

3 Identify the structures on models or charts.

4 Pronounce the name of each term as you point to it.

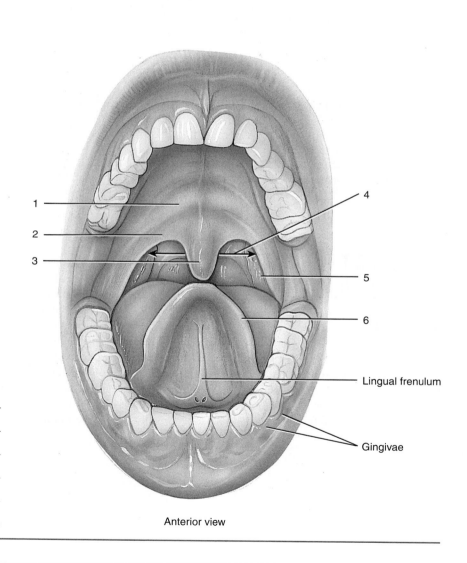

- fauces
- hard palate
- palatine tonsil
- soft palate
- tongue
- uvula

1 _____

2 _____

3 _____

4 _____

5 _____

6 _____

Anterior view

FIGURE 26.3 The mouth.

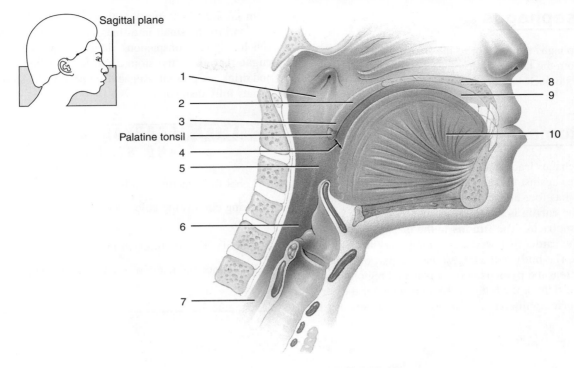

Sagittal plane

Palatine tonsil

Sagittal view

- esophagus
- fauces
- hard palate
- laryngopharynx (la-RIN-go-fair-inks)
- nasopharynx (NA-zo-fair-inks)
- oral or buccal cavity
- oropharynx (OR-o-fair-inks)
- soft palate
- tongue
- uvula (YOU-vu-la)

1 _____

2 _____

3 _____

4 _____

5 _____

6 _____

7 _____

8 _____

9 _____

10 _____

FIGURE 26.4   **Mouth and pharynx.**

3. Esophagus

The esophagus is posterior to the trachea. Food and drink move from the laryngopharynx to the esophagus and then into the stomach.

4. Stomach

The esophagus leads into the superior part of the stomach, which lies against the diaphragm superiorly. The J-shaped stomach has four areas: the cardia, fundus, body, and pylorus. The **cardia** is the first part of the stomach, inferior to the esophagus. The **fundus** is the elevated area to the left of the cardia that serves as a temporary holding area for food. The **body** is the largest, curved part of the stomach, whereas the **pylorus**, a funnel-shaped region, continues to the right of the body of the stomach after the curve. The **pyloric sphincter** is a smooth muscle between the stomach and small intestine that allows food to enter the first part of the small intestine, the duodenum. The stomach has large, conspicuous ridges in the mucosa called **rugae** that allow the stomach to be stretched when food and drink are present. Rugae disappear as the stomach becomes fully distended.

ACTIVITY 4 ESOPHAGUS AND STOMACH

1 Label the structures on Figure 26.5.

2 Using contrasting colors, color the structures in Figure 26.5.

3 Identify the structures on models or charts.

4 Pronounce the name of each term as you point to it.

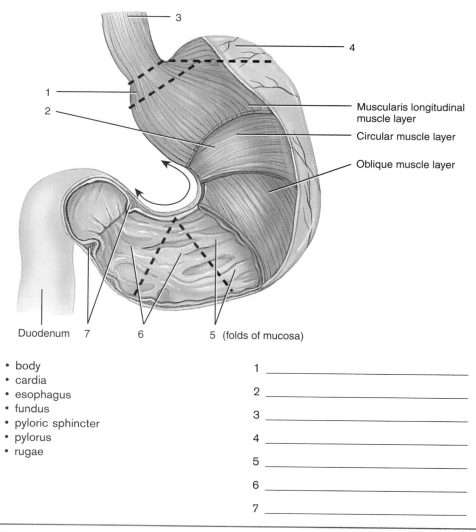

Muscularis longitudinal muscle layer

Circular muscle layer

Oblique muscle layer

Duodenum   7   6   5 (folds of mucosa)

- body
- cardia
- esophagus
- fundus
- pyloric sphincter
- pylorus
- rugae

1 _____

2 _____

3 _____

4 _____

5 _____

6 _____

7 _____

FIGURE 26.5   Esophagus and the stomach.

5. Small Intestine

The small intestine is a long tube (about 10 feet long) divided into three sections: the duodenum, jejunum, and ileum. The shortest section, the **duodenum,** resembles the shape of a "C" and receives food through the pyloric sphincter. The duodenum (*duodenum* = 12 fingers) is a little less than 1 foot long. The next section, the **jejunum** (*jejunum* = empty), is about 3 feet long and has a thicker layer of smooth muscle. It was named jejunum because on death it is found empty. The **ileum** (*ileum* = intestine), the last section of the small intestine, is about 6 feet long and ends at the **ileocecal sphincter** (valve) where it joins the large intestine. From the exterior, it is difficult to distinguish these three sections of the small intestine. The small intestine is designed to be the major site of digestion and absorption.

6. Large Intestine

The large intestine, also known as the colon (*colon* = food passage) or large bowel, is approximately 5 feet long and is composed of the cecum, colon, rectum, and anal canal. A blind pouch called the **cecum,** that is about 2 to 3 inches long, extends inferiorly from the large intestine at the ileocecal sphincter in the lower right quadrant. The slender **appendix,** sometimes called the vermiform appendix (*vermiform* = worm-like; *appendix* = appendage), branches off the cecum and is also about 2 to 3 inches long. From the ileocecal junction, the **ascending colon** extends superiorly to the liver on the right side of the abdomen and makes a 90° turn to the left. It continues as the **transverse colon** until it reaches the spleen, where it makes another 90° turn. The **descending colon** then continues inferiorly on the left side to an S-shaped curve called the **sigmoid colon** at the level of the iliac crest. Bands of smooth muscle on the exterior of the colon contract to form many pouches called **haustra** that give a gathered appearance to the colon. Following the sigmoid colon, at about the 3rd sacral vertebra level, is the **rectum.** The last segment of the colon is the **anal canal** that opens to the exterior at an opening called the **anus.** The anus has two sets of sphincters that control the elimination of feces. The large intestine absorbs a large amount of water, concentrating the undigested feces into a solid form.

CLINICAL NOTE: *A sigmoidoscopy views the lining of the sigmoid colon while a colonoscopy views the lining of the entire colon.*

ACTIVITY 5 SMALL AND LARGE INTESTINES

1 Label the structures on Figures 26.6(a) and (b).

2 Using contrasting colors, color the structures in Figures 26.6(a) and (b).

3 Identify the structures on models or charts.

4 Pronounce the name of each term as you point to it.

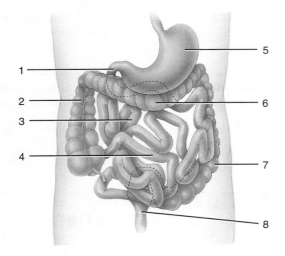

(a) Anterior view

- ascending colon
- descending colon
- duodenum
- ileum
- jejunum
- rectum
- stomach
- transverse colon

1 _____

2 _____

3 _____

4 _____

5 _____

6 _____

7 _____

8 _____

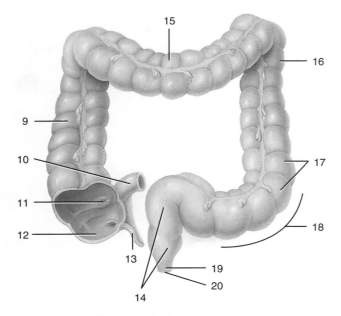

(b) Large intestine, anterior view

- anal canal
- anus
- appendix
- ascending colon
- cecum (SEE-cum)
- descending colon
- haustra (HAW-struh)
- ileocecal sphincter (ileo-SEE-cal)
- ileum
- rectum
- sigmoid colon
- transverse colon

9 _____

10 _____

11 _____

12 _____

13 _____

14 _____

15 _____

16 _____

17 _____

18 _____

19 _____

20 _____

FIGURE 26.6 **Small intestine and large intestine.**

D. ACCESSORY DIGESTIVE ORGANS

1. Salivary Glands

The mucous membranes of the mouth and tongue have many small salivary glands that secrete saliva. However, most of the saliva secreted into the oral cavity is secreted by the parotid, submandibular, and sublingual salivary glands.

The large **parotid** (*par-* = near; *ot-* = ear) **glands** are located anterior and inferior to the ears between the skin and the masseter muscle. The **submandibular glands** are located in the posterior part of the mouth floor just medial to the mandible. The **sublingual glands** are located under the tongue as the name suggests, more medial and superior than the submandibular glands.

2. Tongue

Extrinsic muscles hold the tongue in position but also allow movement to maneuver food in the mouth. The tongue is composed of skeletal muscle called **intrinsic muscles** that are used for speech and swallowing. The **lingual frenulum** attaches the tongue to the floor of the mouth.

3. Teeth

Teeth mechanically break up food into small pieces and are aided by the tongue that manipulates food in the oral cavity.

ACTIVITY 6 SALIVARY GLANDS AND TONGUE

1 Label and study the structures on Figure 26.7.

2 Using contrasting colors, color the structures in Figure 26.7.

3 Identify the structures on models or charts.

4 Pronounce the name of each term as you point to it.

5 Identify incisors, cuspids, premolars, and molars on a skull with teeth.

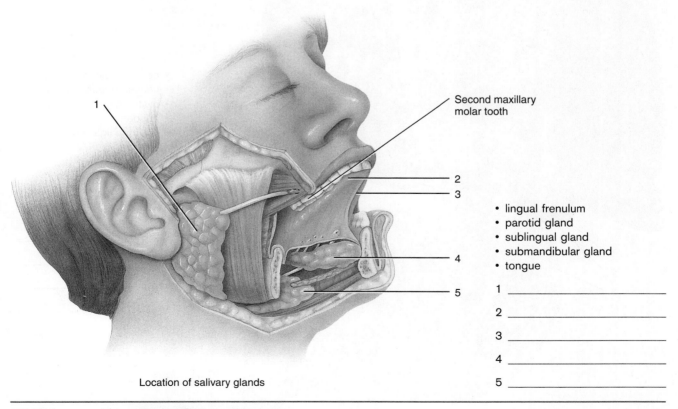

Second maxillary molar tooth

- lingual frenulum
- parotid gland
- sublingual gland
- submandibular gland
- tongue

1 _____

2 _____

3 _____

4 _____

5 _____

Location of salivary glands

FIGURE 26.7   The salivary glands and tongue.

4. Pancreas

The **pancreas** is located posterior to the stomach and lies in the C-shape of the duodenum. Digestive enzymes secreted by the pancreas travel to the duodenum through the **pancreatic duct** which joins the common bile duct to empty into the duodenum.

5. Liver and Gallbladder

The liver is a large organ that lies inferior to the diaphragm and touches the diaphragm along the liver's superior border. The liver has two main lobes, the **right lobe** and **left lobe,** that are separated by the **falciform** (*falc-* = sickle-shaped) **ligament. The common hepatic duct** carries bile to the cystic duct of the gallbladder. The liver makes bile that is delivered to the gallbladder for concentration and storage.

The **gallbladder** is a sac made of smooth muscle that lies on the inferior front part of the liver. Bile leaves the gall-

bladder through the **cystic duct** that joins with the **common hepatic duct** to form the **common bile duct.** The common bile duct joins the pancreatic duct to enter the duodenum. At this site bile salts aid in transforming large fat globules into smaller fat particles for rapid digestion to occur.

ACTIVITY 7 THE PANCREAS, LIVER, AND GALLBLADDER

1 Label and review the structures on Figure 26.8.

2 Using contrasting colors, color the structures in Figure 26.8.

3 Identify the structures on models or charts.

4 Pronounce the name of each term as you point to it.

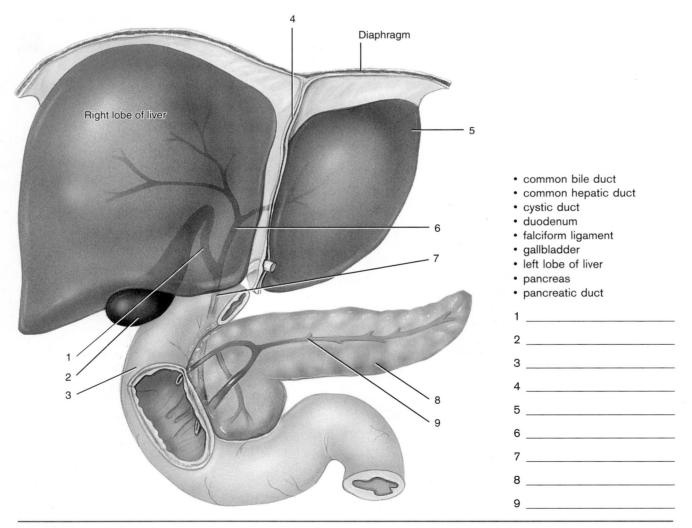

Right lobe of liver

Diaphragm

- common bile duct
- common hepatic duct
- cystic duct
- duodenum
- falciform ligament
- gallbladder
- left lobe of liver
- pancreas
- pancreatic duct

1 _____

2 _____

3 _____

4 _____

5 _____

6 _____

7 _____

8 _____

9 _____

FIGURE 26.8   Pancreas, liver, and gallbladder.

REVIEWING YOUR KNOWLEDGE

A. Gastrointestinal Tract Organs

Write the name of the structure that the phrase describes.

alimentary canal
anus
duodenum
esophagus
large intestine
mouth
pharynx
small intestine
stomach

_____ **1.** Carries food from the pharynx to the stomach.

_____ **2.** A continuous digestive tube from the mouth to the anus.

_____ **3.** Conducts both air and food.

_____ **4.** Receives bile, pancreatic secretions, and food from the stomach.

_____ **5.** Composed of the duodenum, jejunum, and ileum.

_____ **6.** Has two sphincters that control elimination of feces from the body.

_____ **7.** Has regions called the cardia, fundus, body, and pylorus.

_____ **8.** Has regions called the cecum, colon, rectum, and anal canal.

_____ **9.** Receives secretions from salivary glands; mastication occurs here.

B. Digestive Organ Structures

Write the name of the digestive organ structure described below.

cardia
hard palate
haustra
ileocecal sphincter
large intestine
pyloric sphincter
pylorus
rugae
soft palate

_____ **1.** Folds in the gastric mucosa.

_____ **2.** Valve between the stomach and duodenum.

_____ **3.** Constriction that connects the small and large intestine.

_____ **4.** Bony plate between the mouth and nose.

_____ **5.** Keeps food from going up into nasopharynx.

_____ **6.** Region of stomach where the lower esophageal sphincter meets the stomach.

_____ **7.** Has regions called ascending, transverse, descending, and sigmoid colon.

_____ **8.** Funnel-shaped region of the stomach before the small intestine.

_____ **9.** A series of gathered pouches in the large intestine.

C. Accessory Digestive Organs and the Peritoneum

Write the name of the organ that matches the description.

falciform ligament   pancreas
gallbladder   parotid
greater omentum   sublingual
liver   teeth
mesentery   tongue

_____ **1.** Mechanically breaks up food in mastication.

_____ **2.** Secretes digestive enzymes into the duodenum.

_____ **3.** Salivary glands located under the tongue.

_____ **4.** Produces and secretes bile into ducts.

_____ **5.** Peritoneal membrane that holds the small intestine to the posterior abdominal wall.

_____ **6.** Peritoneal fold that holds the liver to the anterior abdominal wall.

_____ **7.** Manipulates food in mastication.

_____ **8.** Fatty, large fold of peritoneum covering the transverse colon and small intestine.

_____ **9.** Largest salivary glands located anterior and inferior to ears.

_____ **10.** Stores and secretes bile into the duodenum.

MECHANICAL AND CHEMICAL DIGESTION

27

OBJECTIVES

After completing this exercise, you should be able to:

- Name the substrate for each of the following enzymes and the products produced from the enzyme-catalyzed reaction: amylase, protease, peptidase, and lipase
- Describe the importance of mechanical and chemical digestion
- Describe the importance of pH on digestive enzyme function
- Describe the action of bile salts on lipids

MATERIALS

- test tubes, clamps or holders, and racks
- wax markers

Starch Digestion

- saltine crackers (starch)
- pH 7.4 buffer in dropper bottle
- salivary amylase in dropper bottle
- 0.5% hydrochloric acid (HCl) in dropper bottle
- Lugol's iodine (IKI) solution in dropper bottle
- clean spot plates (contain depressions)

Sugar Test

- Benedict's reagent in dropper bottle
- 500-mL beakers and hot plate per group

Bile Emulsification

- bile salts
- vegetable oil

A. CHEMICAL DIGESTION

The food we eat contains macromolecules of carbohydrates, lipids, and proteins. Because these large molecules cannot be absorbed by the gastrointestinal tract, enzymes secreted by various parts of the digestive system **catabolize** or break down large molecules into small ones that can be absorbed. **Starch,** a carbohydrate, is digested into **disaccharides** (*di-* = two; *saccharide* = sugar) and **monosaccharides** (*mono-* = one) by the action of specific enzymes called **amylases. Lipids** are macromolecules that are catabolized or digested to **glycerol** and **fatty acids** by the enzyme **lipase. Proteases** digest **proteins** into **peptides** and **amino acids. Peptidases** digest **peptides** into amino acids. Monosaccharides, glycerol and fatty acids, and amino acids are small enough to be absorbed across the

323

wall of the gastrointestinal tract and are used by cells to build new macromolecules or to provide energy (ATP).

Enzymes are typically large protein molecules produced by cells of the body that are special biological catalysts acting on specific **substrates** to produce **products.** Enzymes speed up the rate of chemical reactions between substrates without becoming chemically changed in the reaction. Some enzymes **anabolize** (build) small molecules into macromolecules, whereas others **catabolize** macromolecules like carbohydrates, lipids, and proteins into small molecules. Enzymes work best on food substrates that are chewed and broken down mechanically into smaller particles with a greater surface area.

Each enzyme has optimal environmental conditions, typically temperature and pH, that allow it to be the most active. Enzymes in our body function well at body temperature (37°C) but will denature and become inactive if temperature becomes too high. The pH of the digestive system changes from the mouth to the small intestine; therefore, the enzymes in these different areas have different optimal pHs.

In this exercise, we are using one specific enzyme, salivary amylase, to demonstrate digestion of carbohydrates into glucose. Whole pieces of cracker (starch) are compared with crumbled crackers to observe the importance of mechanical digestion. The activity of amylase will be observed at a pH similar to that of the oral cavity and a pH similar to the stomach to demonstrate the importance of pH on enzyme activity. Carbohydrate digestion begins in the mouth with a pH of about 7.4, but salivary amylase is denatured by the low pH in the stomach from the presence of hydrochloric acid (HCl). Pancreatic buffers bring the pH back to 7.4, a favorable pH for pancreatic amylase, allowing carbohydrate digestion to continue in the small intestine.

SAFETY PRECAUTIONS
- Wear safety glasses when heating test tubes and working with acids.
- Use a test tube clamp for handling heated test tubes.

ACTIVITY 1 DIGESTION OF STARCH

1 Digestion of starch by amylase.
- Obtain 6 test tubes, test tube rack, test tube holder, and a wax marker.
- Obtain 1 saltine cracker and one dropper bottle each of: amylase, 0.5% hydrochloric acid (HCl), and pH 7.4 buffer.
- Mark the tubes 1 through 6 with the wax pencil and put your initials on the upper part of the test tube

for identification in the water bath. Place tubes in the test tube rack. Test tubes 1 and 2 are your controls to be used as comparisons with your results in test tubes 3 through 6.
- Place ⅛ of a piece of a saltine cracker, uncrumbled, in tubes 1, 3, and 5. Add 12 drops of buffer to each tube and swirl it around.
- Crumble ⅛ of a saltine cracker into very small pieces; crush with a mortar and pestle, if available, and place in tubes 2, 4, and 6. Add 12 drops of buffer to each tube and swirl it around.
- Swirl to mix the amylase solution before using. Add 3 drops of *amylase* to tubes 3 through 6.
- Add 3 drops of *0.5% hydrochloric acid (HCl)* to tubes 5 and 6.
- Add enough *water* to tubes 1 through 4 to equal the liquid in tubes 5 and 6.
- Cover the test tubes with Parafilm and shake the tubes well to mix the contents.
- Remove the Parafilm and place the test tube holder with the 6 filled tubes in a 37°C water bath. Incubate for at least 30 minutes.
- Appoint a timer for timing the tubes, and have this person gently shake the test tube rack periodically for best results.
- Proceed to the bile salts preparation setup in Activity 2 while these tubes incubate.

2 Start water boiling in the 500-mL beaker for the sugar test.

3 Determine the amount of starch in each tube.
- Obtain 6 spot plates and 1 dropper bottle each of Lugol's iodine (IKI) solution and Benedict's solution.
- Mark the depressions on the spot plates 1 through 6, matching the test tube identification.
- Obtain test tubes from the water bath and swirl to mix.
- Place 2 drops of each test tube solution in the appropriately marked depression on the spot plate.
- Place one drop of **Lugol's iodine (IKI)** solution (brown) on top of the sample drops.
- Observe the color and record this in Table 27.1. IKI turns **blue** or **black** depending on the amount of starch present and indicates a **positive starch test.** If there is **no color change,** this indicates a **negative starch test.**

brown or color of IKI solution (−)

light blue (+)

medium blue (+ +)

dark blue (+ + +)

black (+ + + +)

4 Determine the amount of **sugar** formed when starch is digested by amylase.

- Place 3 drops of **Benedict's solution** (light blue) in the remaining solutions in all 6 tubes and swirl.

- Put all 6 tubes into the beaker when the water is boiling and time it for 3 minutes.

- Observe the color of the solution and record this in Table 27.1. A **positive sugar test** is green, yellow, orange, or red, depending on the amount of sugar present. If the color remains blue, you have a **negative sugar test.**

  blue (−)

  green (+)

  yellow (+ +)

  orange (+ + +)

  red (+ + + +)

- Compare the results of the whole pieces of crackers in test tubes 1, 3, and 5 to the crumbled crackers in test tubes 2, 4, and 6.

- Clean up your lab table area and return your supplies as directed by your instructor.

5 Answer discussion questions with your lab group.

6 Discuss the results of this exercise with the class.

❓ DISCUSSION QUESTIONS: DIGESTION OF STARCH

1 How did the results of the whole piece of cracker differ from the crumbled pieces of cracker? Explain.

2 Did amylase work better in the pH 7.4 solution or the acidic solution? Explain.

3 Which tubes tested positive for both starch and sugar? Explain.

4 Did any tubes test positive for sugar only? Explain.

5 Did any tubes test negative for sugar? Explain.

6 How would the results be different if you had a longer incubation time? Explain.

7 Which test tube mimicked the conditions in the stomach? Explain.

TABLE 27.1 Digestion of Starch by Salivary Amylase

			TUBE NO.			
	1	**2**	**3**	**4**	**5**	**6**
Contents added to test tubes	Whole cracker Buffer Water	Crumbled cracker Buffer Water	Whole cracker Buffer Amylase Water	Crumbled cracker Buffer Amylase Water	Whole cracker Buffer Amylase 0.5% HCl	Crumbled cracker Buffer Amylase 0.5% HCl
Lugol's iodine (IKI) test results for starch						
Benedict's test results for sugar						

B. BILE EMULSIFICATION

Bile is not a digestive enzyme. **Bile** is composed of a mixture of substances, including **bile salts,** that aid in digestion by physically breaking up large molecules of lipids into smaller molecules with a greater surface area, a process called **emulsification.**

ACTIVITY 2 BILE EMULSIFICATION

1 Observe the ability of bile salts to emulsify lipids.

- Obtain 2 test tubes and a tube rack and mark the tubes 1 and 2.

- To each test tube, add 12 drops of water and 2 drops of vegetable oil (a lipid).

- To test tube 2, add a pinch of bile salts.

- Cover the tubes with Parafilm and shake them to mix well.

- Let the tube contents stand for 15 minutes.

- Describe the contents of the 2 tubes:

Tube 1 _____

Tube 2 _____

2 Answer discussion questions with your group.

❓ DISCUSSION QUESTIONS: BILE EMULSIFICATION

1 Which tube had the oil emulsified? _____

2 Is bile a digestive enzyme? _____

3 Explain how emulsification increases chemical digestion of lipids by lipases.

_____

REVIEWING YOUR KNOWLEDGE

A. General Terminology

Write the term that matches the description.

absorption   macromolecule
amylase   product
enzyme   protease
lipase   substrate

_____ **1.** Large molecules resulting from anabolism.

_____ **2.** A biological catalyst.

_____ **3.** The molecule an enzyme acts on.

_____ **4.** Process that happens to nutrients after digestion.

_____ **5.** Enzyme that hydrolyzes lipid.

_____ **6.** Enzyme that hydrolyzes protein.

_____ **7.** Enzyme that hydrolyzes carbohydrates.

_____ **8.** The molecule produced as a result of an enzymatic process.

B. Substrates and Products

Fill in the blanks.

amino acids   glycerol
emulsification   monosaccharides
fatty acids   mouth

1. Proteins are hydrolyzed into and absorbed as _____.

2. Lipids are hydrolyzed into and absorbed as _____ and _____.

3. Carbohydrates are hydrolyzed into and absorbed as _____.

4. Carbohydrate digestion begins in the _____.

5. Bile salts prepare lipids for digestion by a process called _____.

C. Carbohydrate Digestion Exercise

1. The reagent used to test for the presence of starch was _____.

2. The reagent used to test for the presence of sugar was _____.

3. Three food macromolecules digested in our GI tract are _____, _____, and _____.

4. Why were there 2 tubes in Activity 1 that had no enzyme added to them?

_____

5. The enzymes that digest carbohydrates work best at a(n) _____ pH.

6. The reason why we chew food (and crumbled the cracker in Activity 1) is to _____.

7. A positive test for the presence of sugar is indicated by the colors:

_____

8. A positive test for the presence of starch is indicated by the colors:

_____

STRUCTURE OF THE URINARY SYSTEM AND URINALYSIS

28

OBJECTIVES

After completing this exercise, you should be able to:

- Locate and identify the organs of the urinary system on models or charts
- Describe how urine is formed and trace the path urine takes to the exterior of the body
- Describe the major features of a urinalysis

MATERIALS

- male and female urogenital models or charts
- human kidney models
- fresh or preserved sheep or pig kidney, dissecting tray and tools, disposable gloves
- urinometer, 100-mL beaker, urine test strips, artificial urine obtained from a biological supply house (1 normal and 3 different abnormal urines)
- colored pencils

Urinary system organs include the kidneys, ureters, urinary bladder, and urethra. The kidney regulates water, electrolyte, and pH balance in the body. It removes wastes and nitrogenous substances from blood and excretes them in the urine. Urine is formed in the kidneys and flows through the ureters to the urinary bladder that stores urine until it is eliminated from the body through the urethra.

A. GROSS ANATOMY OF URINARY SYSTEM ORGANS

1. Location and Structure of the Kidneys

Kidneys are bean-shaped structures that are **retroperitoneal** (*retro* = behind), located between the abdominal wall and the peritoneum. They are found at waist level between the 12th thoracic vertebra and the 3rd lumbar vertebra.

The **renal capsule,** a thin fibrous membrane, covers the outer surface of the kidney. The **renal cortex** is the most superficial region of the kidney, and the **renal medulla** is deep to the cortex. The medulla contains cone-shaped **renal pyramids** and extensions of the cortex, **renal columns,** that are found between pyramids. The base of each pyramid faces the cortex, and the **renal papilla** is the apex that is pointed toward the center of the kidney. The cortex and medulla contain **nephrons,** the structural and functional units of the kidney that form urine. Urine drains into ducts that exit through openings of the renal papilla into cup-like **calyces** (calyx, singular), that in turn drain into the **renal pelvis** and the **ureters.**

ACTIVITY 1 LOCATION AND STRUCTURE OF KIDNEYS

1 Label the structures on Figures 28.1 and 28.2.

2 Using contrasting colors, color the urinary structures in Figures 28.1 and 28.2.

3 Locate the kidneys in the superior lumbar area on a human torso.

   a. Which kidney is lower?

   b. What organ forces this kidney into a lower position?

4 Identify the structures from Figure 28.2 on a dissectable kidney model.

5 Pronounce the terms as you point to them.

6 Using an articulated skeleton or a drawing of a skeleton, identify which ribs help protect the kidneys.

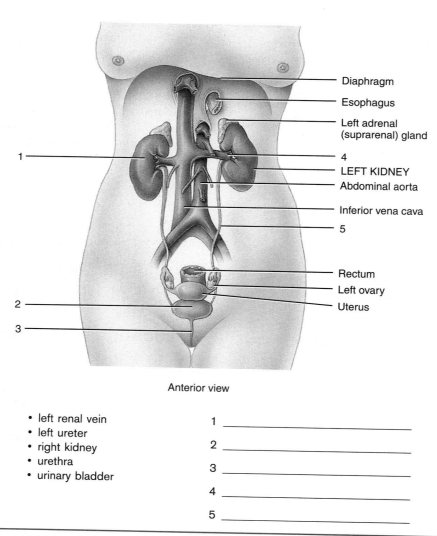

Anterior view

• left renal vein
• left ureter
• right kidney
• urethra
• urinary bladder

1 _____

2 _____

3 _____

4 _____

5 _____

FIGURE 28.1   Organs of the female urinary system in relation to surrounding structures.

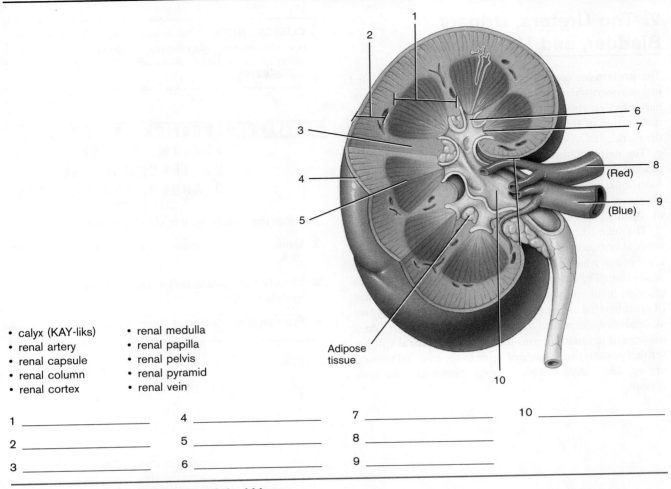

Adipose
tissue

- calyx (KAY-liks)
- renal artery
- renal capsule
- renal column
- renal cortex
- renal medulla
- renal papilla
- renal pelvis
- renal pyramid
- renal vein

1 _____   4 _____   7 _____   10 _____

2 _____   5 _____   8 _____

3 _____   6 _____   9 _____

FIGURE 28.2   Internal structure of the kidney.

2. The Ureters, Urinary Bladder, and Urethra

The **ureters** are narrow, long muscular tubes also located in a retroperitoneal position. They descend toward the urinary bladder, curving medially as they approach the inferior portion of the bladder and enter the posterior wall of the bladder at an oblique angle.

The **urinary bladder** is a hollow, muscular organ that distends to store urine and has three internal openings that form a triangle, the **trigone.** The two posterior openings are the ureteral openings, whereas the anterior opening is the opening into the urethra, the internal urethral orifice.

The tube-like **urethra** carries urine from the internal urethral orifice to the **external urethral orifice,** the opening through which urine exits the body during **micturition** or voiding. The **internal urethral sphincter** is a layer of circular, involuntary smooth muscle that controls passage of urine into the urethra from the urinary bladder. Voluntary skeletal muscle within the urogenital diaphragm (deep muscles of perineum) forms the **external urethral sphincter** that permits the passage of urine to the external urethral orifice. The female urethra is much shorter than the male urethra.

> **CLINICAL NOTE:** *BPH, or benign prostatic hypertrophy, is a noncancerous enlargement of the prostate gland that restricts the urethra and inhibits urine flow, causing urinary retention.*

ACTIVITY 2 LOCATION AND GROSS ANATOMY OF THE URETERS, URINARY BLADDER, AND URETHRA

1 Label the urinary system structures on Figure 28.3.

2 Using contrasting colors, color the structures in Figure 28.3.

3 Identify these structures on male and female urogenital models or charts.

4 Pronounce the names as you point to them.

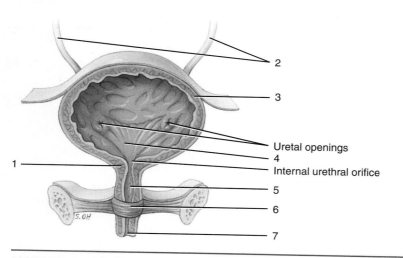

- external urethral orifice
- external urethral sphincter
- internal urethral sphincter
- muscle of urinary bladder
- trigone (TRY-goan)
- ureters
- urethra

1 _____

2 _____

3 _____

4 _____

5 _____

6 _____

7 _____

FIGURE 28.3   Ureters, urinary bladder, and urethra in a female.

B. DISSECTION OF A PIG (OR SHEEP) KIDNEY

Pig and sheep kidneys are very similar to human kidneys. This dissection will illustrate the location and structure of the calyces (plural of calyx) and renal pelvis deep within the kidney.

ACTIVITY 3 KIDNEY DISSECTION

1 Identify the bolded structures listed in this activity on a pig or sheep kidney using Figure 28.4 as a reference.

2 Protect your hands with disposable gloves.

3 Obtain a fresh or preserved kidney, dissecting tray, and scalpel or knife.

4 If you have a fresh kidney, carefully remove any fat on its exterior. Fresh kidneys may have adrenal glands located on the superior surface. Adrenal glands are usually not present on preserved kidneys.

5 Identify the **ureter, renal artery,** and **renal vein** that travel through the hilus. The renal artery will have a thicker wall than the renal vein.

6 Using a scalpel or knife, carefully cut along the frontal plane to separate the kidney into anterior and posterior sections.

7 Observe the cut surface. Using a needle probe, pull the thin **renal capsule** away from the surface of the kidney. The outer region of the kidney, the **cortex,** will be a slightly lighter color than the deeper **medulla.**

8 Blood vessels may be observed in preserved kidneys that are injected with red and blue latex and will appear as red and blue dots within the cortex.

9 Within the medulla, extensions of light-colored cortical tissue, **renal columns,** are found between the darker, cone-shaped **renal pyramids.** Identify the **renal papillae** located at the apices of the renal pyramids.

10 The urine collecting areas are deep to the renal papillae. Identify the **calyces** and the **renal pelvis.**

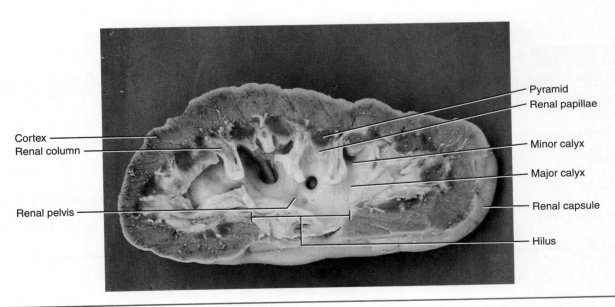

FIGURE 28.4   **Frontal section of preserved pig kidney.**

C. URINALYSIS

A **urinalysis** is an analysis of the physical, chemical, and microscopic characteristics of urine and a measure of urine volume. The normal characteristics of urine are listed in Table 28.1. **Urine volume** varies depending on the water content of the body, and decreases when body fluid volume is low. The **specific gravity of urine** is the weight of a volume of urine divided by the weight of the same volume of distilled water. Urine weight per volume is higher than distilled water because of the presence of solutes in urine. The more solutes present in urine, the higher the specific gravity. Normal urine contains 95% water and 5% solutes. The solutes found in normal urine include **electrolytes** (sodium, potassium, chloride, and other ions), **urea** (formed from breakdown of amino acids), **creatinine** (formed from breakdown of creatine phosphate in muscle tissue), **uric acid** (formed from breakdown of nucleic acids), and **metabolic end products** of hormones and other substances. Although drugs are not a normal solute in urine, they will be excreted in urine if present in the bloodstream.

Selected abnormal components of urine are listed in Table 28.2. The presence of these substances in urine indicates an abnormality in metabolism or kidney function.

ACTIVITY 4 URINALYSIS

1 Obtain four 100-mL beakers and mark them to match the labels on the four urine samples.

2 Obtain 50 mL from each of the four artificial urine samples and transfer into the appropriate beaker. Record the color and turbidity in Table 28.3.

3 *Optional:* Collect and test your own urine sample. Collect your urine in a specimen container supplied by your instructor.

4 Obtain a urinometer and measure the specific gravity for each sample using instructions provided by your instructor. Do not discard urine; use it for the next procedure.

5 Using a urine test strip, measure pH, presence or absence of glucose, albumin, blood, and ketone bodies. Urine test strips contain squares of different reagent paper that change color when they contact specific reagents (chemicals). Record the results in Table 28.3.

6 Compare your results with the rest of the class.

TABLE 28.1 Normal Characteristics of Urine

CHARACTERISTIC	DESCRIPTION
Volume	One to two liters per 24 hours; considerable variation in normal volume
Color	Yellow or amber color; urine color is darker in concentrated urine
Turbidity	Transparent in freshly voided urine; will turn cloudy after standing; microbes, pus, epithelial cells, or crystals may cause cloudiness in fresh urine
Odor	Aromatic when fresh; ammonia-like after standing because of breakdown of urea to ammonia by bacteria
pH	Normal range is 4.6–8.0, with an average of 6.0; high protein diets produce an acidic urine; vegetarian diets produce an alkaline urine
Specific gravity	Normal range is 1.001–1.035; low specific gravity represents dilute urine, higher values represent a concentrated urine.

TABLE 28.2 Selected Abnormal Constituents of Urine

ABNORMAL CONSTITUENTS	DISORDER AND CAUSE
Glucose	**Glucosuria** (glucose in urine) is usually caused by diabetes mellitus but sometimes may be caused by stress (epinephrine stimulates glycogen breakdown); diabetes mellitus and stress result in high blood glucose levels and therefore high levels of glucose in the filtrate; glucose transporters cannot work fast enough to reabsorb all glucose from filtrate.
Red blood cells	**Hematuria** (erythrocytes in urine) may be caused by inflammation of urinary system organs, irritation by kidney stones, kidney disease, trauma to urinary system organs, or polyps or tumors within the urinary system.
White blood cells	**Pyuria** (white blood cells in urine) is caused by an infection in the kidney, ureters, urinary bladder, or urethra.
Albumin	**Albuminuria** (excess albumin in urine) is from an increase in filtration membrane permeability caused by high blood pressure, kidney trauma, disease, or inflammation; trace amounts of albumin in urine is normal.
Ketone bodies	**Ketonuria** (ketone bodies in urine) is caused by a metabolic condition called ketosis; ketosis occurs when cells do not have enough glucose to completely break down fatty acids; ketosis can occur from starvation, low carbohydrate diets, or untreated diabetes mellitus.
Casts	**Casts** are hardened clumps of material formed by protein precipitation and/or cell agglutination within renal tubules; the following conditions may result in cast formation: albuminuria, abnormally acidic urine, and highly concentrated urine.
Calculi	**Calculi** are insoluble salt crystals that can form anywhere within the kidney tubules, ureters, urinary bladder, or urethra; commonly called kidney stones, calculi can cause considerable pain as they pass through the lumens of urinary system organs.
Microbes	Presence of bacteria or other microorganisms indicate the presence of a **urinary tract infection (UTI);** normal urine is sterile (no microorganisms present), however microbes present on the skin surrounding the external urethral orifice may contaminate the urine sample if the urine sample is not carefully obtained.

TABLE 28.3 Urinalysis Results

TEST	SAMPLE A	SAMPLE B	SAMPLE C	SAMPLE D
Color				
Turbidity				
Specific gravity				
pH				
Glucose				
Albumin				
Blood				
Ketone bodies				

REVIEWING YOUR KNOWLEDGE

A. URINARY SYSTEM STRUCTURES

Write the renal structure that the phrase describes.

calyx
nephron
renal capsule
renal columns
renal cortex
renal medulla
renal papilla
renal pelvis
renal pyramids

_____ **1.** Urine-forming structure of the kidney.

_____ **2.** Region of kidney deep to cortex.

_____ **3.** Extensions of renal cortex found in between renal pyramids.

_____ **4.** Apex of renal pyramid.

_____ **5.** Covers outer surface of kidney.

_____ **6.** Receives urine from several calyces.

_____ **7.** Cup-like structure that receives urine from renal papilla.

_____ **8.** Cone-shaped structures located within the renal medulla.

_____ **9.** Outermost region of the kidney, just deep to the capsule.

B. Ureters, Urinary Bladder, and Urethra

Write the name of the structure(s) that the phrase describes.

external urethral orifice   trigone
external urethral sphincter   ureters
internal urethral sphincter   urinary bladder
micturition

_____ **1.** Voiding of urinary bladder.

_____ **2.** Stores urine until it is voided.

_____ **3.** Area bounded by ureteral openings and internal urethral orifice.

_____ **4.** Voluntary skeletal muscle in urogenital diaphragm that allows passage of urine to exterior of body.

_____ **5.** Carries urine from renal pelvis to urinary bladder.

_____ **6.** Urine is excreted through this opening.

_____ **7.** Circular smooth muscle that involuntarily controls passage of urine from the urinary bladder to the urethra.

C. Urine Formation and Flow

Trace the flow of filtrate and urine through the urinary system, writing the structures in order.

calyces
nephron
renal papillae
renal pelvis
ureters
urethra
urinary bladder

1. _____

2. _____

3. _____

4. _____

5. _____

6. _____

7. _____

MALE AND FEMALE REPRODUCTIVE SYSTEMS

OBJECTIVES

After completing this exercise, you should be able to:

- Describe the structure and function of the male reproductive organs

- Identify the location of male reproductive organs on urogenital models or charts

- Trace the pathway of a male gamete from its production through the duct system to ejaculation

- Describe the structure and function of the female reproductive organs

- Identify the location of the female reproductive organs on models or charts

- Trace the pathway of a female gamete from its production to fertilization and trace the pathway of a zygote from formation to implantation in the uterus

MATERIALS

- dissectible models or charts of the male reproductive system

- dissectible models or charts of the female reproductive system

- colored pencils

A. GROSS ANATOMY OF THE MALE REPRODUCTIVE SYSTEM

The male reproductive system has **testes** that produce **gametes** called **sperm (spermatozoa)** and male sex hormones. Ducts store and transport the sperm, and accessory sex glands secrete fluid that, along with sperm, form **semen.** Supporting reproductive structures are the **scrotum** that houses the testes and the **penis** that transfers sperm into the female vagina.

1. Testes and Sperm

The **testes (testicles** or **male gonads)** are oval-shaped glands that are covered and protected by the scrotum (scrotal sac). The **scrotum** is a pouch of loose skin located on the exterior of the body that provides a temperature 2 to 3°C lower than body temperature that is needed for normal sperm development. The testes produce and release testosterone and sperm. **Sperm** are produced in tiny **seminiferous tubules** located within lobules in the testes.

Sperm are also called **sperm cells** or **spermatozoa.** Mature sperm have lost most of their cytoplasm and

339

cellular organelles, and do not survive in the female reproductive tract more than a few days after ejaculation. Sperm are composed of a head, midpiece, and tail. The **head** consists mostly of the nucleus with an acrosome (*acro-* = top) covering it. The **acrosome** is a vesicle filled with enzymes that allow sperm to penetrate the female oocyte for fertilization. The **midpiece** contains many mitochondria that are needed to produce energy (ATP) for movement of the tail once the sperm cells are ejaculated. The **tail** is a flagellum that performs a whip-like motion, propelling the sperm.

2. Duct System

After leaving the **seminiferous tubules,** sperm travel through a series of tubes that empty into the **epididymis** (*epi-* = on or upon; *didymus* = pair), a curved, tubular organ located on the posterior side of each testis. The epididymis stores sperm as they undergo maturation. Peristaltic contractions of smooth muscle in the epididymis move sperm into the **ductus deferens** (*duct* = to lead; *deferens* = carrying away) or **vas deferens** (*vas* = vessel). The ductus deferens leaves the scrotum and enters the **spermatic cord** that travels between the testes and the inguinal canal. Also within the spermatic cord are spermatic arteries, veins, and nerves. The ductus deferens enters the anterior abdominal cavity through the **inguinal** (*inguinal* = groin) **canal** and loops around the posterior of the urinary bladder to enter the **prostate gland** posteriorly. Each ductus deferens joins the duct from the seminal vesicle to form an ejaculatory duct. The **ejaculatory ducts**, located in the prostate gland, empty sperm and fluid into the **urethra.** Semen travels through the urethra and is ejaculated from the penis through the **external urethral orifice.**

3. Accessory Sex Glands and Semen

Three types of accessory glands—the seminal vesicles, prostate gland, and bulbourethral (Cowper's) glands—make and secrete most of the fluid portion of the **semen** (*semen* = seed). The testes secrete only a small amount of fluid before the addition of fluid from these three glands.

Seminal vesicles (*seminal* = pertaining to semen) are sac-like glands posterior to the bladder, flanking either side. When sperm cells reach the end of the ductus deferens, the duct of the seminal vesicle joins it and adds seminal fluid. The chestnut-size **prostate gland** (*prostate* = standing before) surrounds the prostatic urethra and adds prostatic fluid to the semen. The smallest accessory glands are the paired, pea-sized **bulbourethral (Cowper's) glands** (*bulbus* = swollen root) that are located within the urogenital diaphragm and add fluid that lubricates the penile urethra and glans penis.

4. Penis

The **penis** is a cylindrical shaft medial and anterior to the scrotum and contains the urethra. The penis is composed of three main parts: a body, a root, and the glans penis. The body of the penis contains three separate cylinders of erectile tissue, all containing blood sinuses. Fibrous connective tissue, fascia, and skin surround all three cylinders.

The **glans penis** (*glans* = acorn) is the distal, expanded portion of the penis that is mostly covered with **prepuce** (foreskin) in an uncircumcised male. **Circumcision** removes the prepuce, exposing the glans penis. The **external urethral orifice** is the external opening of the urethra in the glans penis.

ACTIVITY 1 STRUCTURE OF THE MALE REPRODUCTIVE SYSTEM

1 Label the structures on Figures 29.1, 29.2, and 29.3.

2 Using contrasting colors, color the structures in Figures 29.1, 29.2, and 29.3.

3 Locate the structures on models or charts.

4 Pronounce the terms as you point to them.

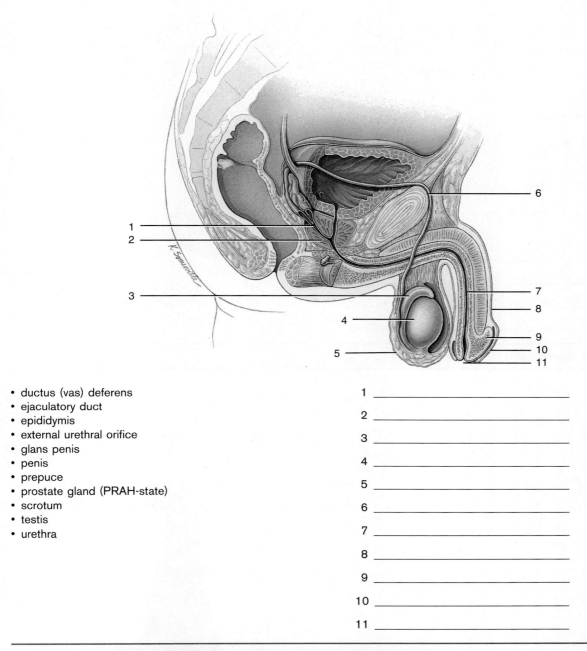

- ductus (vas) deferens
- ejaculatory duct
- epididymis
- external urethral orifice
- glans penis
- penis
- prepuce
- prostate gland (PRAH-state)
- scrotum
- testis
- urethra

1 _____

2 _____

3 _____

4 _____

5 _____

6 _____

7 _____

8 _____

9 _____

10 _____

11 _____

FIGURE 29.1   Male reproductive organs and duct system, sagittal section.

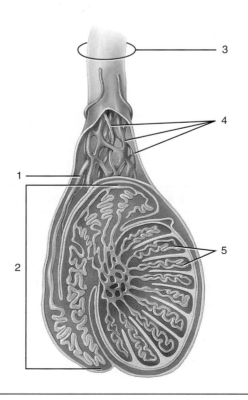

- acrosome (AK-row-sohm)
- head
- midpiece
- mitochondria
- nucleus
- tail (flagellum)

1 _____

2 _____

3 _____

4 _____

5 _____

6 _____

FIGURE 29.2   Sperm cell.

- blood vessels and nerves
- epididymis (ep-ih-DID-ih-miss)
- ductus (vas) deferens (DUK-tus DEAF-er-ens)
- seminiferous tubules (sem-ih-NIF-er-us)
- spermatic cord (sper-MAT-ic)

1 _____

2 _____

3 _____

4 _____

5 _____

FIGURE 29.3   The epididymis and ductus deferens, sagittal section.

B. GROSS ANATOMY OF THE FEMALE REPRODUCTIVE SYSTEM

The female reproductive system has ovaries that produce gametes (oocytes) and female sex hormones, tubes to transport the oocytes, and a uterus to protect and nourish the developing embryo and fetus. In addition, the female reproductive system has accessory structures: the vagina, external genitalia, and mammary glands.

1. Ovaries and Uterine Tubes

The paired, oval-shaped **gonads** or **ovaries** (*ovary* = egg receptacles) are the size of unshelled almonds (about an inch long) and are located on either side of the uterus. The ovaries are anchored in place by three paired ligaments: the broad ligaments, the ovarian ligaments, and the suspensory ligaments. **Oocytes** (immature eggs or gametes) are produced in the ovary and mature within **ovarian follicles,** and are released during a process called ovulation. Oocytes become **ova** (mature eggs) after fertilization.

Paired **uterine (Fallopian) tubes** or **oviducts** curve around the ovaries to receive and transport the oocyte to the superior part of the uterus. The finger-like structures projecting from the expanded open end of the tube are called **fimbriae.** During ovulation the beating of the fimbriae forms a current that sweeps the oocyte into the uterine tube. Fertilization of the oocyte by a sperm cell usually occurs here. The uterine tube attaches to the superior, lateral part of the uterus.

2. Uterus and Vagina

The pear-shaped **uterus** is a little smaller than the size of your fist, and is posterior to the urinary bladder and anterior to the rectum. Externally, three main regions of the uterus can be seen: the fundus, body, and cervix. The **fundus** is the dome-shaped superior portion of the uterus that is wider and meets the entrance of the uterine tubes on both sides of the uterus. The **body** starts inferior to the openings of the uterine tubes, is the largest region of the uterus, and decreases in size to meet the narrower cervix inferiorly. The space within the uterus is the **uterine cavity.** The narrowed **cervix** opens into the vagina inferiorly. The uterine wall has three layers: an outer **perimetrium,** a middle **myometrium,** and an inner **endometrium.** The myometrium is the muscular layer that forms the majority of the uterine wall. Endometrial secretions will nourish the zygote once conception takes place. Part of the endometrial layer is shed during menstruation.

Although the uterus is secured in its position by four paired ligaments, only two of these ligaments are easily found on models: the large, flat broad ligaments and the cord-like round ligaments.

The **vagina** (*vagina* = sheath) is a cylindrical passageway for sexual intercourse, childbirth, and menstrual flow. Inferior to the cervix of the uterus, the muscular vagina is angled posteriorly and is located between the urinary bladder anteriorly and the rectum posteriorly. The vagina opens to the exterior through the **vaginal orifice.**

3. External Genitalia

The external female genitalia are referred to as the **vulva** (*vulva* = to wrap around) and consist of the mons pubis, labia majora, labia minora, clitoris, and vestibule. The **mons** (*mons* = mountain) **pubis** is an anterior pad of adipose tissue beneath the skin that covers the pubic symphysis. Posterior to this are two longitudinal folds of adipose tissue covered with skin and pubic hair called the **labia majora** (singular, labium majus). Medial to these larger folds are smaller paired longitudinal folds of skin without pubic hair called the **labia minora** (singular, labium minus). The **clitoris,** a cylindrical erectile tissue, is posterior to the mons pubis and anterior to the external urethral orifice. The **prepuce,** an anterior extension of the labia minora, covers the body of the clitoris like a hood, but leaves part of the clitoris exposed. The **vestibule** is the area medial to the paired labia minora and contains the external urethral orifice, vaginal orifice, hymen (if present), and openings of glands that secrete mucus for lubrication. The **hymen** (*hymen* = membrane) is a thin membrane around the perimeter of the vaginal orifice that partially blocks the orifice.

ACTIVITY 3 THE STRUCTURE OF THE FEMALE REPRODUCTIVE ORGANS

1 Label the structures on Figures 29.4 and 29.5.

2 Using contrasting colors, color the structures in Figures 29.4 and 29.5.

3 Identify the structures on urogenital models or charts.

4 Pronounce the terms as you point to them.

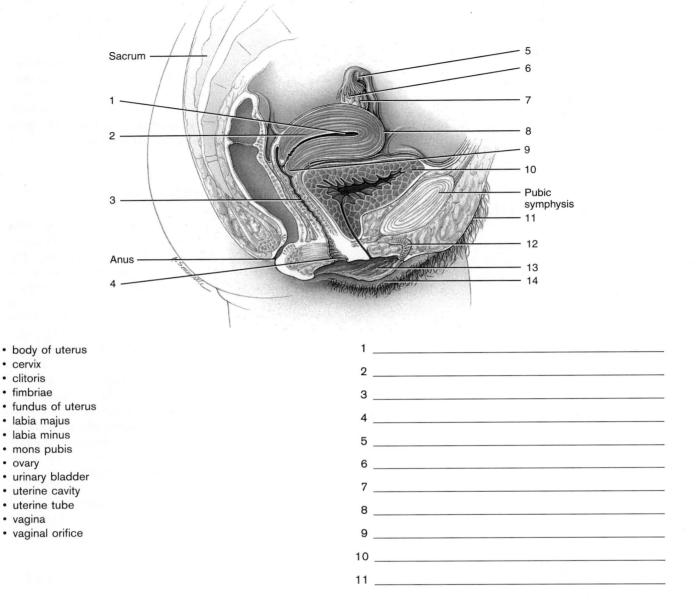

Sacrum

1

2

3

Anus

4

5

6

7

8

9

10

Pubic symphysis

11

12

13

14

- body of uterus
- cervix
- clitoris
- fimbriae
- fundus of uterus
- labia majus
- labia minus
- mons pubis
- ovary
- urinary bladder
- uterine cavity
- uterine tube
- vagina
- vaginal orifice

1 _____

2 _____

3 _____

4 _____

5 _____

6 _____

7 _____

8 _____

9 _____

10 _____

11 _____

12 _____

13 _____

14 _____

FIGURE 29.4 **Female reproductive organs, sagittal section.**

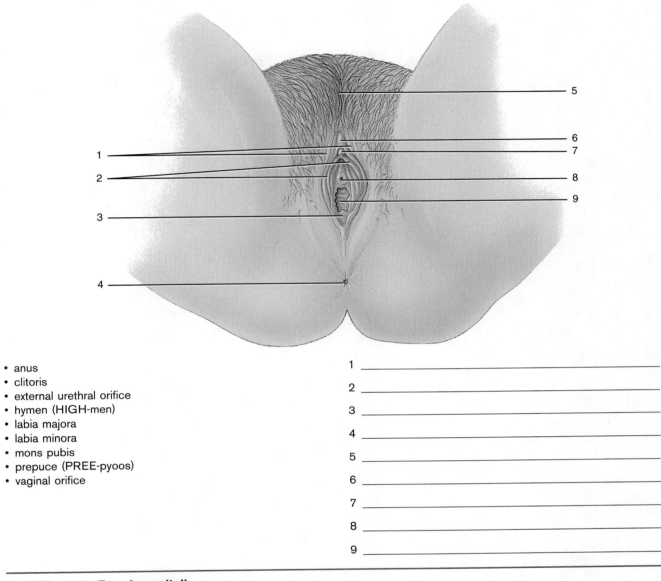

- anus
- clitoris
- external urethral orifice
- hymen (HIGH-men)
- labia majora
- labia minora
- mons pubis
- prepuce (PREE-pyoos)
- vaginal orifice

1 _____

2 _____

3 _____

4 _____

5 _____

6 _____

7 _____

8 _____

9 _____

FIGURE 29.5   Female genitalia.

REVIEWING YOUR KNOWLEDGE

A. Male Reproductive System

Write the name of the structure that the phrase describes.

bulbourethral glands   prepuce   spermatic
cord
ejaculatory duct   prostate gland   testis
epididymis   scrotum   vas deferens
penis   seminal vesicles

_____ 1. Encompasses the ductus deferens, spermatic blood vessels, and nerves.

_____ 2. Organ that produces sperm.

_____ 3. Site of storage and maturation of sperm.

_____ 4. Located in prostate; union of ducts of the ductus deferens and seminal vesicle.

_____ 5. Copulatory organ; passageway for urine and semen.

_____ 6. Removed during circumcision.

_____ 7. Covers and protects testes.

_____ 8.
                            ⎫
_____ 9. ⎬ Three glands that produce seminal fluid.
                            ⎪
_____ 10.⎭

_____ 11. Tube located in both the spermatic cord and pelvic cavity for transporting sperm.

B. Male Duct System

List the following ducts in order, numbering them 1 through 5, tracing sperm from their production to ejaculation.

_____ ductus (vas) deferens

_____ epididymis

_____ ejaculatory duct

_____ seminiferous tubules

_____ urethra

C. Female Reproductive System

Write the name of the structure that the phrase describes.

adipose tissue ovary
body of uterus uterine tube
clitoris uterus
cervix vagina
fimbriae vestibule
fundus of uterus vulva
labia majora

_____ **1.** Domed portion of uterus.

_____ **2.** Transports oocyte to uterus.

_____ **3.** Produces oocytes and hormones.

_____ **4.** Finger-like projections on end of uterine tubes.

_____ **5.** Narrower portion of uterus; connects with vagina.

_____ **6.** Largest portion of uterus.

_____ **7.** Implantation of zygote and development of fetus occur in this organ.

_____ **8.** Area between the labia minora with openings of urethra and vagina.

_____ **9.** Another name for external genitalia.

_____ **10.** Erectile tissue of female.

_____ **11.** Tissue that pads mons pubis.

_____ **12.** Larger longitudinal folds of skin and pubic hair.

_____ **13.** Organ for intercourse; canal for childbirth and menstruation.

D. Pathway of a Female Gamete

Number the structures in order, 1 through 5, showing the pathway of a gamete from the ovary to the uterus.

_____ body of uterus

_____ fimbriae of uterine tube

_____ fundus of uterus

_____ ovary

_____ uterine tube

ANSWER KEY

Exercise 1

Figure 1.1

1. skull
2. face
3. armpit
4. forearm
5. wrist
6. thigh
7. ankle
8. chest
9. arm
10. abdomen
11. groin
12. pelvis
13. foot
14. shoulder
15. back
16. hand
17. buttock
18. leg
19. head
20. neck

Figure 1.2

1. cranial
2. facial
3. cervical
4. acromial
5. axillary
6. brachial
7. antecubital
8. carpal
9. digital
10. femoral
11. patellar

12. tarsal
13. digital
14. frontal
15. orbital
16. otic
17. buccal
18. nasal
19. oral
20. sternal
21. pectoral
22. umbilical
23. coxal
24. inguinal
25. pubic
26. pedal
27. occipital
28. scapular
29. vertebral
30. olecranal
31. gluteal
32. popliteal
33. fibular (peroneal)
34. plantar
35. lumbar
36. calcaneal

Activity 4

a. anterior
b. superior
c. medial
d. distal
e. ipsilateral
f. superficial

Figure 1.3

1. posterior
2. superior

3. anterior
4. proximal
5. distal
6. inferior

Figure 1.4

1. frontal or coronal section
2. sagittal section
3. transverse section

Exercise 2

Figure 2.1

1. digestive
2. nervous
3. endocrine
4. lymphatic and immune
5. integumentary
6. muscular
7. cardiovascular
8. respiratory
9. reproductive
10. skeletal
11. urinary

Activity 2

1. a
2. b
3. d
4. c
5. d
6. b
7. b
8. a
9. b
10. b

Figure 2.2

1. cranial cavity
2. vertebral canal
3. dorsal cavity
4. thoracic cavity
5. diaphragm
6. abdominal cavity
7. pelvic cavity
8. abdominopelvic cavity
9. ventral cavity
10. pleural cavity
11. pericardial cavity
12. thoracic cavity
13. diaphragm
14. abdominal cavity
15. brim of pelvis
16. pelvic cavity
17. abdominopelvic cavity

Table 2.2

1. brain
2. spinal cord
3. lungs
4. heart
5–14. stomach, pancreas, liver, gall bladder, small intestine, large intestine, appendix, kidneys, spleen, adrenal glands
15–21. urinary bladder, large intestine, rectum, ovaries, uterus, vagina, prostate gland

Activity 4

a. RLQ
b. all four quadrants
c. mainly RUQ; some LUQ
d. LLQ and RLQ
e. mainly LUQ
f. all four quadrants
g. LUQ
h. mainly LUQ

Exercise 3

Activity 2

10. The letter e is upside down and a mirror image to the initial drawing.
11. When the stage is moved to the right, the object viewed through the lenses appears to be moving to the left. When the stage is moved away from the observer, the object viewed through the lenses appears to be moving toward the observer.

Activity 4 Discussion Question: Cheek Smear

1. The cell is clear, addition of stain allows cell structure to be observed.

Exercise 4

Activity 1 Discussion Questions: Cell Specialization

1. a. long, cylindrical; nucleus
   b. columnar; cilia and nucleus
   c. irregular shape with processes that extend from cell body; nucleus
   d. oval cell; nucleus and flagellum
   e. disc; none
   f. round; nucleus
2. size, shape, and organelles present

Figure 4.3

a. interphase
b. prophase
c. metaphase
d. anaphase
e. telophase

Exercise 5

Activity 1 Discussion Questions: Diffusion

1. potassium permanganate
2. it has smaller molecules

Activity 2

1. isotonic; hypertonic; hypotonic
2. not change shape; swell; crenate

Figure 5.2

a. normal shape, isotonic solution
b. swollen, hypotonic solution
c. crenated, hypertonic solution

Activity 2 Discussion Questions: Osmosis in RBCs

1. no
2. yes
3. water would enter the RBCs, but slower than if the cells are in distilled water
4. hypertonic and hypotonic solutions

Activity 3

2. Copper sulfate will pass through the filter. Charcoal will not pass through the filter.

Activity 3 Discussion Questions: Filtration

1. plasma membrane of a cell
2. hydrostatic pressure
3. yes; filtration was faster at the beginning when hydrostatic pressure was higher

Exercise 6

Figure 6.3

1. nucleus of connective tissue cell
2. connective tissue
3. simple squamous epithelium
4. nucleus of simple squamous epithelial cell

Figure 6.4

1. simple cuboidal epithelium
2. apical surface of simple cuboidal epithelial cell
3. lumen of kidney tubule
4. nucleus of simple cuboidal epithelial cell

Answer to question: The tubules are not straight, but contain many curves. Parts of a tubule may be perpendicular to plane of section (appear circular in section), parallel to plane of section (long oval in section), or oblique to plane of section (short oval).

Figure 6.5

1. microvilli on apical surface of simple columnar epithelial cell
2. nucleus of simple columnar epithelial cell
3. simple columnar epithelium
4. connective tissue

Figure 6.6

1. nucleus of squamous epithelial cell
2. stratified squamous epithelium
3. nucleus of epithelial cell in basal layer of epithelium
4. connective tissue

Figure 6.7

1. nucleus of transitional epithelial cell in apical layer
2. nucleus of transitional epithelial cell in basal layer
3. transitional epithelium
4. connective tissue

Figure 6.8

1. nucleus of ciliated columnar epithelial cell
2. cilia on apical surface of columnar epithelial cell
3. pseudostratified ciliated columnar epithelium
4. connective tissue

Figure 6.10

1. collagen fiber
2. elastic fiber
3. connective tissue cells (fibroblasts or lymphocytes)

Figure 6.11

1. reticular fiber
2. reticulocyte

Figure 6.12

1. nucleus of adipocyte
2. lipid storage area

Figure 6.13

1. fibroblast
2. collagen fiber bundle

Figure 6.14

1. fibroblast
2. collagen fiber bundles running in different directions
3. parallel collagen fiber bundles

Figure 6.15

1. fibroblast
2. elastic fiber

Figure 6.16

1. extracellular matrix
2. lacuna
3. nucleus of chondrocyte

Figure 6.17

1. lamella
2. canaliculus
3. lacuna
4. central canal

Figure 6.18

1. red blood cells
2. nucleus of white blood cell
3. platelet

Figure 6.20

1. width of individual muscle fiber
2. nucleus
3. striation

Figure 6.21

1. width of cardiac muscle fiber
2. nucleus
3. branches of cardiac muscle fiber
4. intercalated discs

Figure 6.22

1. nucleus of smooth muscle fiber in cross-section
2. nucleus of smooth muscle fiber in longitudinal section

Figure 6.24

1. processes
2. cell body of multipolar neuron
3. nucleus

Exercise 7

Figure 7.1

1. dermal papillae
2. epidermis
3. dermis
4. hypodermis

Figure 7.2

1. stratum corneum
2. stratum basale

Figure 7.3

1. hair shaft
2. hair root
3. sebaceous gland
4. arrector pili muscle
5. hair follicle
6. hair bulb
7. eccrine sweat gland
8. apocrine sweat gland

Figure 7.4

1. sebaceous gland
2. hair follicle
3. hair root
4. hair bulb

Activity 2 Discussion Questions: Eccrine Gland Density

1. Right palm and forehead had the highest density; right anterior forearm the lowest.
2. Areas with greatest density of eccrine sweat glands will have the most sweat during exercise or nervousness.

Exercise 8

Figure 8.1

a. short
b. flat
c. irregular
d. sesamoid
e. long

Figure 8.2

1. proximal epiphysis
2. diaphysis
3. distal epiphysis
4. articular cartilage
5. epiphyseal line
6. compact bone
7. medullary cavity
8. yellow marrow
9. periosteum
10. endosteum
11. spongy bone

Activity 3

2. The articular cartilage is shiny and the periosteum is dull.

Activity 4

1. a. not flexible at all
   b. more brittle
   c. collagen/organic matter
   d. increases risk of bone breakage
2. a. increased flexibility
   b. not hard; very soft
   c. calcium phosphate/minerals/salts
   d. rickets

Exercise 9

Figure 9.1

1. parietal bone
2. coronal suture
3. squamous suture

4. temporal bone
5. lambdoid suture
6. external auditory meatus
7. occipital bone
8. mastoid process
9. frontal bone
10. sphenoid bone
11. lacrimal bone
12. nasal bone
13. zygomatic bone
14. condylar process of mandible
15. maxilla
16. mandible

Figure 9.2

1. maxilla
2. zygomatic arch
3. vomer
4. sphenoid bone
5. jugular formen
6. foramen magnum
7. occipital bone
8. zygomatic bone
9. palatine bone
10. mandibular fossa
11. carotid foramen
12. styloid process
13. occipital condyle
14. temporal bone

Figure 9.3

1. parietal bone
2. occipital bone
3. frontal bone
4. crista galli
5. cribriform plate
6. ethmoid bone
7. optic foramen
8. sella turica
9. sphenoid bone
10. temporal bone

Figure 9.4

1. sagittal suture
2. parietal bone
3. ethmoid bone
4. lacrimal bone
5. perpendicular plate of ethmoid
6. inferior nasal concha
7. vomer
8. frontal bone
9. squamous suture
10. temporal bone
11. nasal bone
12. middle nasal concha of ethmoid
13. zygomatic bone
14. maxilla
15. mandible

Figure 9.5

1. frontal sinus
2. ethmoidal sinus
3. sphenoidal sinus
4. maxillary sinus

Figure 9.6

1. anterior (frontal) fontanel
2. posterior (occipital) fontanel
3. posterior (occipital) fontanel
4. anterior (frontal) fontanel

Figure 9.7

1. cervical curve
2. thoracic curve
3. lumbar curve
4. sacral curve
5. cervical vertebrae
6. thoracic vertebrae
7. lumbar vertebrae
8. sacrum
9. coccyx

Figure 9.9

1. superior articular facet
2. transverse foramen
3. transverse process
4. dens (odontoid process)
5. spinous process
6. body
7. transverse process
8. bifurcated spinous process

Figure 9.10

1. spinous process
2. transverse process
3. vertebral foramen
4. body
5. facet on transverse process
6. slanted spinous process
7. body

Figure 9.11

1. spinous process
2. transverse process
3. vertebral foramen
4. body
5. hatchet-shaped spinous process
6. body

Figure 9.12

1. intervertebral foramen
2. inner pulp-like center
3. outer fibrous part
4. intervertebral disk
5. herniation
6. inner pulp-like center
7. outer fibrous part

Figure 9.13

1. coccyx
2. sacral canal
3. auricular surface (for sacroiliac joint)
4. sacral foramen

Figure 9.14

a. scoliosis
b. kyphosis
c. lordosis

Figure 9.15

1. manubrium
2. body of sternum
3. xiphoid process
4. sternum
5. costal cartilage
6. true ribs
7. floating ribs
8. false ribs

E x e r c i s e   1 0

Figure 10.1

1. acromion
2. coracoid process
3. glenoid cavity
4. acromion
5. spine of scapula
6. glenoid cavity

Figure 10.2

1. capitulum
2. head
3. deltoid tuberosity
4. coronoid fossa
5. trochlea
6. olecranon fossa

Figure 10.3

1. head of radius
2. styloid process of radius
3. olecranon (process)
4. trochlear notch (semilunar)
5. coronoid process
6. styloid process of ulna
7. ulna
8. radius

Figure 10.4

1. carpals
2. metacarpals
3. 5th proximal phalanx
4. 5th middle phalanx
5. 5th distal phalanx

Figure 10.5

1. ilium
2. greater sciatic notch
3. ischium
4. ischial spine
5. obturator foramen
6. ischial tuberosity
7. iliac crest
8. anterior superior iliac spine
9. acetabulum
10. pubis

Figure 10.7

1. head of femur
2. greater trochanter
3. neck
4. medial condyle
5. lateral condyle

Figure 10.8

1. lateral condyle
2. head of fibula
3. fibula
4. lateral malleolus
5. medial condyle
6. tibial tuberosity
7. tibia
8. medial malleolus

Figure 10.9

1. calcaneus
2. talus
3. tarsals
4. metatarsals
5. 2nd proximal phalanx
6. 2nd middle phalanx
7. 2nd distal phalanx
8. phalanges

E x e r c i s e   1 1

Figure 11.1

1. synovial
2. synovial
3. cartilaginous
4. synovial
5. fibrous
6. fibrous
7. cartilaginous
8. cartilaginous
9. synovial
10. synovial
11. fibrous
12. fibrous

Figure 11.2

1. synarthrosis
2. amphiarthrosis
3. diarthrosis
4. amphiarthrosis
5. diarthrosis
6. diarthrosis
7. diarthrosis
8. diarthrosis
9. amphiarthrosis
10. synarthrosis

Figure 11.3

1. articular bone
2. synovial fluid
3. synovial cavity
4. articular cartilage
5. fibrous capsule
6. synovial membrane
7. articular capsule

Figure 11.4

1. flexion
2. extension
3. hyperextension
4. hyperextension
5. extension
6. flexion

Figure 11.5

1. abduction
2. adduction
3. circumduction
4. adduction
5. abduction

Figure 11.6

1. rotation
2. lateral rotation
3. medial rotation

Figure 11.7

1. inversion
2. eversion
3. dorsiflexion
4. plantar flexion
5. pronation
6. supination

Exercise 12

Figure 12.1

1. endomysium
2. skeletal muscle fiber in cross-section
3. fascicle

4. perimysium
5. width of skeletal muscle fiber
6. nucleus
7. striation

Figure 12.2

1. axon terminal
2. synaptic end bulb
3. synaptic vesicle with neurotransmitter
4. motor end plate

Figure 12.3

1. skeletal muscle fibers
2. axon terminal with synaptic end bulbs
3. motor neuron axon

Activity 2 Discussion Questions: Observation of Unfused Tetanus, Recruitment, and Fatigue

1. Incomplete tetanus and asynchronous contractions allow you to maintain a contraction over time.
2. Motor unit recruitment. More motor units recruited, more cross-bridges attached, and more force generated.
3. More motor units contracting at a given time; as motor units fatigue, less motor units to take over. Therefore it takes less time to fatigue all motor units.

Exercise 13

Activity 1

1. orbicularis—direction of fibers (circular); oculi—location (eye)
2. biceps—number of origins (two); brachii—location (arm)
3. rectus—orientation of muscle (muscle parallel to body midline); femoris—location (thigh)
4. gluteus—location (buttocks); maximus—muscle size (largest)
5. deltoid—shape (triangular)
6. sternocleidomastoid—origin and insertion (sterno- and cleido- indicate origin on sternum and clavicle; mastoid indicates insertion on mastoid process)

Figure 13.1

1. frontalis
2. orbicularis oculi
3. zygomaticus major
4. orbicularis oris
5. frontalis
6. temporalis
7. occipitalis
8. orbicularis oculi
9. masseter
10. zygomaticus major
11. sternocleidomastoid
12. orbicularis oris
13. trapezius
14. platysma

Figure 13.2

1. deltoid
2. pectoralis major
3. serratus anterior
4. external oblique
5. internal oblique
6. transverse abdominis
7. rectus abdominis

Figure 13.3

1. deltoid
2. trapezius
3. latissimus dorsi

Figure 13.4

1. biceps brachii
2. brachialis
3. triceps brachii, lateral head
4. triceps brachii, long head
5. triceps brachii, medial head

Figure 13.5

1. brachioradialis
2. flexor carpi radialis
3. flexor carpi ulnaris
4. flexors
5. flexor carpi ulnaris
6. extensor carpi radialis longus
7. extensor carpi radialis brevis
8. extensor carpi ulnaris
9. extensor digitorum
10. extensors

Figure 13.6

1. sartorius
2. rectus femoris
3. tensor fasciae latae
4. vastus intermedius
5. vastus medialis
6. vastus lateralis
7. pectineus
8. gracilis
9. adductor longus
10. adductor magnus
11. adductors
12. gluteus maximus
13. gluteus medius
14. semitendinosus
15. semimembranosus
16. biceps femoris

Figure 13.7

1. extensor digitorum longus
2. tibialis anterior
3. fibularis (peroneus) longus
4. gastrocnemius
5. soleus
6. flexor digitorum longus

7. tibialis anterior
8. extensor digitorum longus
9. fibularis (peroneus) longus
10. soleus
11. gastrocnemius
12. gastrocnemius
13. soleus
14. calcaneal tendon (Achilles)

Exercise 14

Figure 14.1

1. pia mater
2. dura mater
3. adipose tissue in epidural space
4. subarachnoid space
5. arachnoid mater with web-like projection

Figure 14.2

1. conus medullaris
2. cauda equina
3. filum terminale
4. cervical nerves
5. thoracic nerves
6. lumbar nerves
7. sacral nerves
8. coccygeal nerve

Figure 14.3

1. anterior (ventral) root
2. gray matter
3. white matter
4. posterior (dorsal) root
5. central canal
6. posterior (dorsal) root ganglion
7. spinal nerve

Figure 14.4

1. posterior (dorsal) root
2. white matter
3. central canal
4. gray matter

Exercise 15

Figure 15.1

1. integrating center
2. sensory neuron
3. sensory receptor
4. effector
5. motor neuron

Exercise 16

Figure 16.1

1. cerebellum
2. cerebrum
3. pineal gland
4. thalamus
5. hypothalamus
6. diencephalon
7. midbrain
8. pons
9. medulla oblongata
10. brain stem

Figure 16.2

1. pons
2. medulla oblongata
3. midbrain
4. cerebellum

Figure 16.3

1. frontal lobe
2. temporal lobe
3. parietal lobe
4. occipital lobe

Figure 16.4

1. frontal lobe
2. longitudinal fissure
3. precentral gyrus
4. central sulcus
5. postcentral gyrus
6. parietal lobe
7. gyrus
8. sulcus
9. longitudinal fissure

Figure 16.5

1. superior sagittal sinus
2. arachnoid villus
3. dura mater
4. parietal bone
5. arachnoid mater
6. subarachnoid space
7. pia mater
8. cerebral cortex (gray matter)
9. white matter

Figure 16.6

1. lateral ventricles
2. third ventricle
3. cerebral aqueduct
4. fourth ventricle
5. central canal of spinal cord
6. superior sagittal sinus
7. arachnoid villus
8. subarachnoid space
9. lateral ventricle
10. choroid plexus
11. third ventricle
12. cerebral aqueduct
13. fourth ventricle
14. central canal

Figure 16.7

1. abducens
2. facial
3. vestibulocochlear
4. glossopharyngeal
5. vagus
6. olfactory tract
7. optic
8. oculomotor
9. trochlear
10. trigeminal
11. hypoglossal
12. accessory

Exercise 17

Figure 17.1

1. lacrimal gland
2. lacrimal canals
3. nasolacrimal duct

Figure 17.2

1. anterior cavity
2. scleral venous sinus
3. cilary body
4. vitreous chamber
5. choroid
6. sclera
7. retina
8. cornea
9. pupil
10. iris
11. lens
12. optic nerve

Exercise 18

Figure 18.1

1. auricle
2. external ear
3. malleus
4. incus
5. stapes attached to oval window
6. middle ear
7. semicircular canals
8. vestibule
9. cochlea
10. internal ear
11. auditory tube

12. tympanic membrane
13. external auditory canal

Exercise 19

Figure 19.1

1. pineal gland
2. parathyroid glands
3. adrenal glands
4. pancreas
5. hypothalamus
6. pituitary gland
7. thyroid gland
8. thymus
9. ovaries
10. testes

Figure 19.2

1. posterior pituitary (neurohypophysis)
2. hypothalamus
3. infundibulum
4. anterior pituitary (adenohypophysis)

Figure 19.3

1. right lobe of thyroid gland
2. left lobe of thyroid gland
3. isthmus
4. left parathyroid glands
5. thyroid gland
6. right parathyroid glands

Figure 19.4

1. right adrenal gland
2. left adrenal gland
3. kidney
4. capsule
5. adrenal cortex
6. adrenal medulla

Figure 19.5

1. tail of pancreas
2. body of pancreas
3. head of pancreas

Exercise 20

Table 20.2

Antigens on RBCs: A; B; A and B; none
Antibodies in plasma: B; A; none; A and B
Compatible donor blood types: A and O; B and O; AB, A, B, and O; O
Incompatible donor blood types: B and AB; A and AB; none; A, B, and AB

Table 20.4

Neutrophils: 47.7%
Lymphocytes: 41.0%
Monocytes: 6.5%
Eosinophils: 4.3%
Basophils: 0.5%
% Hematocrit: 41.6
Hb or HGB content g/100 mL: 14.4

Table 20.5

Hematocrit Tube a: RBC column = 44 mm; whole column = 97 mm; hematocrit = 45%
Hematocrit Tube b: RBC column = 20 mm; whole column = 95 mm; hematocrit = 21%

Activity 3 Discussion Questions: Blood Tests

1. Neutrophils are low and lymphocytes are high; long-term infection.
2. Tube (a) = normal; Tube (b) = anemia
3. 2.9:1; Yes
4. Clotting time would increase.
5. Hemophiliacs have higher than normal clotting time.

Exercise 21

Figure 21.1

1. base of heart
2. right lung
3. serous pericardium
4. left lung
5. apex of heart
6. fibrous pericardium
7. diaphragm
8. fibrous pericardium
9. parietal layer of serous pericardium
10. pericardial cavity
11. visceral layer of serous pericardium (epicardium)
12. endocardium
13. myocardium

Figure 21.2

1. superior vena cava
2. right pulmonary artery
3. ascending aorta
4. pulmonary trunk
5. right pulmonary veins
6. right coronary artery
7. auricle of right atria
8. right ventricle
9. inferior vena cava
10. aortic arch
11. ligamentum arteriosum
12. left pulmonary artery
13. left pulmonary veins
14. auricle of left atrium
15. branch of left coronary artery
16. left ventricle

17. descending aorta
18. aortic arch
19. ligamentum arteriosum
20. left pulmonary artery
21. left pulmonary veins
22. coronary sinus
23. superior vena cava
24. ascending aorta
25. right pulmonary artery
26. right pulmonary veins
27. inferior vena cava

Figure 21.3

1. pulmonary (semilunar) valve
2. right atrium
3. coronary sinus opening
4. tricuspid valve
5. right ventricle
6. left atrium
7. aortic (semilunar) valve
8. bicuspid valve (mitral)
9. chordae tendineae
10. interventricular septum
11. papillary muscle
12. left ventricle

Activity 4

2. Trace blood flow

   right atrium
   tricuspid valve
   right ventricle
   pulmonary (semilunar) valve
   pulmonary trunk
   right and left pulmonary arteries
   pulmonary capillaries
   pulmonary veins
   left atrium
   bicuspid valve
   left ventricle
   aortic (semilunar) valve
   aorta
   systemic arteries
   systemic capillaries
   systemic veins
   venae cavae

4. oxygen-rich blood
   aorta
   pulmonary arteries—oxygen-poor
   pulmonary trunk—oxygen-poor
   pulmonary veins—oxygen-rich
   venae cavae—oxygen-poor

Activity 5

8. The three cusps are half-moon-shaped and have no chordae tendineae.
10. Two main cusps that are counted; yes; yes; lesser number; left ventricle is thicker to pump blood through the systemic circulation

Exercise 22

Activity 1 Discussion Questions: Heart Sounds, Heart Rate, and Pulse Rate

1. Lubb; it occurs when the ventricles are contracting.
2. They should be the same.
3. The pause is shorter after exercise.
4. Differences in recovery time may be due to age, health, and physical condition.

Activity 3 Discussion Questions: Regulation of Blood Pressure

1. Supine position. Head is not higher than heart, it does not take as much pressure to drive blood to head. Pressure increases when standing to provide enough pressure to overcome gravity and deliver blood to brain.
2. BP and heart rate are increased. BP was changed by increasing cardiac output.

Exercise 23

Figure 23.1

b. vasoconstricted

Figure 23.2

1. capillary
2. arteriole

Figure 23.3

1. lumen of artery
2. internal layer of artery
3. middle layer of artery
4. outer layer of artery
5. outer layer of vein
6. middle layer of vein
7. inner layer of vein
8. lumen of vein

Figure 23.4

1. left common carotid
2. left subclavian
3. aortic arch
4. left coronary

Figure 23.5

1. internal carotid
2. vertebral
3. common carotid
4. internal carotid
5. vertebral

Figure 23.6

1. subclavian
2. axillary

3. brachial
4. radial
5. ulnar

Figure 23.7

1. celiac trunk
2. common iliac
3. thoracic aorta
4. renal
5. abdominal aorta

Figure 23.8

1. common iliac
2. external iliac
3. femoral
4. popliteal
5. anterior tibial
6. posterior tibial
7. dorsal artery of the foot (dorsalis pedis)

Figure 23.9

1. superior vena cava
2. inferior vena cava

Figure 23.10

1. vertebral
2. internal jugular
3. subclavian
4. brachiocephalic

Figure 23.11

1. subclavian
2. axillary
3. brachial veins
4. radial veins
5. median cubital
6. ulnar veins

Figure 23.12

1. right brachiocephalic
2. left brachiocephalic
3. superior vena cava
4. left renal
5. inferior vena cava
6. left common iliac
7. left external iliac

Figure 23.13

1. hepatic portal
2. superior mesenteric
3. splenic

Figure 23.14

1. external iliac
2. femoral
3. anterior tibial veins
4. femoral

5. popliteal
6. posterior tibial veins

Figure 23.15

1. left pulmonary artery
2. pulmonary trunk
3. left pulmonary vein

Figure 23.16

1. ductus arteriosus
2. foramen ovale
3. ductus venosus
4. umbilical vein
5. umbilical arteries

Exercise 24

Figure 24.1

1. pharyngeal tonsil
2. internal naris
3. opening of auditory tube
4. soft plate
5. uvula
6. palatine tonsil
7. lingual tonsil
8. nasal chonchae or turbinates
9. nasal meatuses
10. external naris
11. hard palate
12. nasopharynx
13. oropharynx
14. laryngopharynx

Figure 24.2

1. epiglottis
2. thyroid cartilage
3. cricoid cartilage
4. vocal folds
5. glottis

Figure 24.3

1. apex
2. visceral pleura
3. parietal pleura
4. pleural cavity
5. superior lobe
6. middle lobe
7. inferior lobe
8. base
9. larynx
10. trachea
11. primary bronchus
12. secondary bronchus
13. tertiary bronchus
14. diaphragm

Figure 24.4

1. respiratory bronchiole
2. alveolar ducts
3. alveolar sac
4. alveoli

Exercise 25

Figure 25.1

1. external intercostals
2. internal intercostals
3. rectus abdominis
4. external oblique
5. internal oblique
6. transverse abdominis
7. diaphragm

Figure 25.2

1. parietal pleura and thoracic wall
2. trachea
3. bronchus
4. lungs covered with visceral pleura
5. pleural cavity
6. diaphragm

Table 25.1

1. decreased
2. increased
3. deflated
4. out of balloons
5. increased
6. decreased
7. inflated
8. into balloons

Activity 2 Discussion Questions: Bell Jar Demonstration

1. serous fluid; larger than pleural cavity
2. (a) subatmospheric (If rubber membrane is not sealed to bell jar, then pressure inside bell jar might be atmospheric.)
(b) subatmospheric
3. two separate pleural cavities; yes; the lung model has only one pleural cavity
4. thorax increases in size with inspiration and decreases with expiration

Exercise 26

Figure 26.1

1. salivary glands
2. pharynx
3. esophagus
4. liver
5. gallbladder
6. large intestine

7. mouth
8. teeth
9. tongue
10. stomach
11. pancreas
12. small intestine
13. anus

Figure 26.2

1. falciform ligament
2. greater omentum
3. greater omentum (reflected)
4. mesentery

Figure 26.3

1. hard palate
2. soft palate
3. uvula
4. fauces
5. palatine tonsil
6. tongue

Figure 26.4

1. nasopharynx
2. soft palate
3. uvula
4. fauces
5. oropharynx
6. laryngopharynx
7. esophagus
8. hard palate
9. oral cavity
10. tongue

Figure 26.5

1. cardia
2. body
3. esophagus
4. fundus
5. rugae
6. pylorus
7. pyloric sphincter

Figure 26.6

1. duodenum
2. ascending colon
3. jejunum
4. ileum
5. stomach
6. transverse colon
7. descending colon
8. rectum
9. ascending colon
10. ileum
11. ileocecal sphincter
12. cecum
13. appendix
14. rectum

15. transverse colon
16. descending colon
17. haustra
18. sigmoid colon
19. anal canal
20. anus

Figure 26.7

1. parotid gland
2. tongue
3. lingual frenulum
4. sublinguinal gland
5. submandibular gland

Figure 26.8

1. cystic duct
2. gallbladder
3. duodenum
4. falciform ligament
5. left lobe of liver
6. common hepatic duct
7. common bile duct
8. pancreas
9. pancreatic duct

Exercise 27

Activity 1 Discussion Questions: Digestion of Starch

1. Less sugar produced by whole cracker. Whole cracker had smaller surface area for amylase to act on.
2. 7.4. Acidic pH denatured the enzyme and made it inactive.
3. Tubes with amylase at pH 7.4 (tubes 3 and 4). Amylase broke down some but not all the starch to sugar.
4. No, not all starch was converted to sugar.
5. Yes, the tubes without amylase (tubes 1 and 2) and the tubes with amylase and HCl (tubes 5 and 6). Without active amylyase, starch could not be broken down to sugar.
6. More sugar would appear in tubes with amylase at pH 7.4.
7. The tube with crumbled cracker, amylase, and HCl (tube 4). The stomach receives masticated food from the mouth and it has an acidic pH that denatures amylase.

Activity 2 Discussion Questions: Bile Emulsification

1. Tube 2
2. No, it doesn't break down fat. It suspends it in the watery medium so it can be broken down by water-soluble lipase.
3. Increases the surface area available for enzymes to digest fats.

Exercise 28

Figure 28.1

1. right kidney
2. urinary bladder

3. urethra
4. left renal vein
5. left ureter

Figure 28.2

1. renal medulla
2. renal cortex
3. renal column
4. renal capsule
5. renal pyramid
6. calyx
7. renal papilla
8. renal artery
9. renal vein
10. renal pelvis

Activity 1

3. a. right kidney
   b. liver

Figure 28.3

1. internal urethral sphincter
2. ureters
3. muscle of urinary bladder
4. trigone
5. urethra
6. external urethral sphincter
7. external urethral orifice

Exercise 29

Figure 29.1

1. ejaculatory duct
2. prostate gland
3. epididymis
4. testis
5. scrotum
6. ductus (vas) deferens
7. urethra
8. penis
9. glans penis
10. prepuce
11. external urethral orifice

Figure 29.2

1. acrosome
2. nucleus
3. mitochondria
4. head
5. midpiece
6. tail (flagellum)

Figure 29.3

1. ductus (vas) deferens
2. epididymis
3. spermatic cord

4. blood vessels and nerves
5. seminiferous tubules

Figure 29.4

1. uterine cavity
2. body of uterus
3. vaginal canal
4. vaginal orifice
5. uterine tube
6. fimbriae
7. ovary
8. fundus of uterus
9. cervix
10. urinary bladder
11. mons pubis
12. clitoris
13. labia minus
14. labia majus

Figure 29.5

1. labia majora
2. labia minora
3. hymen
4. anus
5. mons pubis
6. prepuce
7. clitoris
8. external urethral orifice
9. vaginal orifice

EXERCISE 1: Figure 1-6: Mark Nielsen.

EXERCISE 3: Figure 3-1: Courtesy Olympus America, Inc. Figure 3-2: Bruce Iverson/Photo Researchers.

EXERCISE 4: All photos in Chapter 4 are courtesy Michael Ross, University of Florida, with the exception of Figure 4-2c, d, and e. Figure 4-2c: Carolina Biological/Visuals Unlimited. Figure 4-2d: M. Abbey/Photo Researchers. Figure 4-2e: Michael Abbey/Visuals Unlimited.

EXERCISE 5: Figure 5-2: David Phillips/Photo Researchers.

EXERCISE 6: Figures 6-3 to 6-8, 6-10 to 6-18, and 6-20 to 6-22: Courtesy Michael Ross, University of Florida. Figure 6-24: Jan Leesma/Custom Medical Stock Photo, Inc. Figure 6-25 (#1): Biophoto Associates/Photo Researchers. Figure 6-25 (#2, #3, #4): Ed Reschke. Figure 6-26 (#1, #2, #3, #9, #11, #13, #14, #15): Biophoto Associates/Photo Researchers. Figure 6-26 (#4, #5, #8, #10, #12): Ed Reschke. Figure 6-26 (#6): Courtesy Michael Ross, Univer-

sity of Florida. Figure 6-26 (#7): Courtesy Andrew J. Kuntzman. Figure 6-27(#1, #2): Courtesy Michael Ross, University of Florida. Figure 6-27 (#3): Ed Reshke.

EXERCISE 7: Figures 7-1, 7-2, and 7-4: Courtesy Michael Ross, University of Florida.

EXERCISE 8: Figure 8-3a: Mehau Kulyk/Photo Researchers. Figure 8-3b: Scott Camazine & Sue Trainor/Photo Researchers. Figure 8-4: P. Motta/Photo Researchers.

EXERCISE 9: Figures 9-1 to 9-4, 9-7, 9-9 to 9-11, 9-13, and 9-15: Mark Nielsen.

EXERCISE 10: Figures 10-1 to 10-5, and 10-7 to 10-9: Mark Nielsen.

EXERCISE 11: Figures 11-4 to 11-7: John Wilson White.

EXERCISE 12: Figure 12-1 and 12-3: Courtesy Michael Ross, University of Florida.

EXERCISE 14: Figure 14-7: Mark Nielsen.

EXERCISE 16: Figures 16-1 to 16-3a: Fred Hossler/Visuals Unlimited. Figure 16-7:

Ralph Hutchings/Imagingbody.com. Figures 16-8 to 16-10: Mark Nielsen.

EXERCISE 17: Figure 17-3: Courtesy William Radke, University of Central Oklahoma.

EXERCISE 20: Figure 20-2: From Lennart Nilsson, *Our Body Victorious,* Boehringer Ingelheim International GmbH. Reproduced with permission. Figure 20-3: All photos courtesy Michael Ross, University of Florida, with the exception of Figure 20-3c (#4), 20-3d (#4) and 20-3e (#3), are courtesy John D. Cunningham/Visuals Unlimited. Figure 20-5: JC REVY/ISM/Phototake. Figure 20-6: James Hayden/Phototake.

EXERCISE 21: Figure 21-5: Mark Nielsen.

EXERCISE 23: Figure 23-1: From Phelps, P.C., and J. H. Luft. *Am. J. Anat.* 125:399, 1969. Figure 23-2: T. Kuwabara-D. Fawcett/Visuals Unlimited. Figure 23-3: Carolina Biological/Visuals Unlimited.

EXERCISE 28: Figure 28-4: Courtesy Dr. William Benyak.